Ergonomics Principles in Design

Ergonomics Principles in Design: An Illustrated Fundamental Approach touches upon different ergonomic principles in design and then showcases examples of where and how they have been applied. Each chapter covers one aspect of design and emphasizes its application in the real world, such as the ergonomic design of the interface of a blood pressure monitor and the ergonomic interface of a moving ticket vending machine.

- Discusses all aspects of design from product, space, and communication
- Includes many self-explanatory assignments for better understanding
- Highlights practice sessions at the end of each chapter with design directions to help the readers
- Demonstrates ergonomics principles with the help of real-life examples
- Focuses on the application of ergonomic principles in design in the form of studio assignments

The text covers the application of ergonomic principles in diverse areas of design, like product, space, and communication, in a single volume. It will serve as an ideal reference text for graduate students and professionals in the fields of ergonomics, design, human factors, occupational health and safety, and industrial and manufacturing engineering.

Ergonomics Principles in Design

An Illustrated Fundamental Approach

Prabir Mukhopadhyay, MSc, PhD

CRC Press is an imprint of the
Taylor & Francis Group, an **informa** business

First edition published 2022
by CRC Press
6000 Broken Sound Parkway NW, Suite 300, Boca Raton, FL 33487-2742

and by CRC Press
4 Park Square, Milton Park, Abingdon, Oxon, OX14 4RN

CRC Press is an imprint of Taylor & Francis Group, LLC

Library of Congress Cataloging–in–Publication Data

Names: Mukhopadhyay, Prabir, 1969- author.
Title: Ergonomics principles in design : an illustrated fundamental approach / Prabir Mukhopadhyay.
Description: First edition. | Boca Raton, FL : CRC Press, 2023. | Includes bibliographical references and index.
Identifiers: LCCN 2022012249 (print) | LCCN 2022012250 (ebook) | ISBN 9781032299617 (hbk) | ISBN 9781032299600 (pbk) | ISBN 9781003302933 (ebk)
Subjects: LCSH: Design--Human factors. | Human engineering.
Classification: LCC T59.7 .M8525 2023 (print) | LCC T59.7 (ebook) | DDC 620.8/2--dc23/eng/20220701
LC record available at https://lccn.loc.gov/2022012249
LC ebook record available at https://lccn.loc.gov/2022012250

ISBN: 9781032299617 (hbk)
ISBN: 9781032299600 (pbk)
ISBN: 9781003302933 (ebk)

DOI: 10.1201/9781003302933

Typeset in Sabon
by Deanta Global Publishing Services, Chennai, India

To the Lotus Feet of Sri Ramakrishna Paramhansha Dev
for whom this journey was possible, Pronam to you.

Late Dhirendra Nath Mukhopadhyay (my father)
and Mrs Meena Mukhopadhyay (my mother)

&

Dada, Mamoni, Duggi, and Leto ...

Contents

Preface

This book is a practical book on the application of ergonomic principles in various areas of design, like product, space, and communication. The book touches upon different ergonomic principles in design and then showcases examples of where and how it has been applied. Each chapter of the book touches one aspect of design and emphasizes its application in the real world. The chapters in the book are sequenced in a manner that they become gradually complex as the reader moves from the first to the last chapter.

As designers or design students come from varied backgrounds like arts, commerce, film, science, and technology, it's important that the book is written in a clear and easy-to-understand "layperson's" language avoiding scientific jargon. The illustrations and photographs in the examples make the chapters interesting and hold the reader to the book and its chapters. The anthropometric dimensions used in this book are fictitious and are for demonstration purpose only. These anthropometric dimensions should not be used for making any product for actual use for any specific population. The dimensions have been used to show the different calculations.

The biggest challenge today is that there are no books on the application of ergonomic principles in design for doing studio assignments. This book tries to bridge that gap and explains to those without any background in ergonomics and science the principles of ergonomics and how they could be applied to design. The book can be used by the design students, practitioners, or those in other disciplines also for applying ergonomic principles to different designs of products, space, and communication around them.

Portfolio design is an integral part of the design profession. This book also addresses that aspect. The examples have been chosen in a manner that they are easy to understand and apply. Each of the examples has been carefully selected so that they touch all the aspects of the application of ergonomics in design, and this helps the students and practitioners develop their design portfolios by incorporating the projects on ergonomics.

Each chapter of this book starts with an overview, which lays the foundation for different ergonomic principles that have been applied in the chapter. Every chapter has a practice session, which helps the readers to rehearse whatever he/she has learned in the chapter. The practice sessions also have some design directions for helping the readers in their practice sessions.

Again I wish to express my gratitude to those people who have helped in preparing this book. I am particularly grateful to the staff of CRC Press, Taylor & Francis Group, for their continued editorial and production support all through the publication process.

Acknowledgments

Acknowledgment is due to many people for helping me in completing this book.

Ms Neha Mandlik, Faculty of Furniture and Interior Design, National Institute of Design, Ahmedabad, India, for the following illustrations:

All illustrations for Chapters 1, 2, 3, 8, and 9.

Illustrations for the following chapters:

Chapter 4: Figures 4.1, 4.2, 4.13, 4.15, 4.16, 4.20, 4.21, 4.24, and 4.26, 4.27, 4.28, 4.29, 4.30.

Chapter 5: Figures 5.6, 5.7, 5.13–5.16, 5.19, and 5.21.

Chapter 7: Figure 7.1.

The following students have worked hard toward data collection, analysis, and concept generation, and I owe them a lot.

Chapter 4: Ajay Boga and Ankur Mehra, Post-Graduate Students of Furniture and Interior Design, National Institute of Design, Ahmedabad, India.

Chapter 5: Ramit Kumar Chowdhury, Sushant Saini, Syamlal K, Post-Graduate Student, Transportation and Automobile Design, National Institute of Design, Gandhinagar, India.

Chapter 6: Example of fire extinguisher label and way finding: Joyal KS and Vipin Toshniwal, Master of Design Student, Design Discipline, Indian Institute of Information Technology Design and Manufacturing Jabalpur, India.

Example of pesticide packaging: Rohit Singh, Master of Design Student, Design Discipline, Indian Institute of Information Technology Design and Manufacturing Jabalpur, India.

Chapter 7: Example 1: Sonal Nigam, Post-Graduate Student, Software and User Interface Design, National Institute of Design, Gandhinagar, India.

Example 2: Divesh Jaiswal, Post-Graduate Student, New Media Design, National Institute of Design, Gandhinagar, India.

Example 3: Mayunkhini Pandey, Post-Graduate Students, Strategic Design Management, National Institute of Design, Gandhinagar, India.

A very special thanks to Mr Vipul Vinzuda, Faculty of Transportation and Automobile Design, Post-Graduate Campus, National Institute of Design, Gandhinagar, India, for inspiring me to write this book and helping me fine-tune many of the figures and especially for contributing to Figures 9.3–9.8 at very short notice.

Author

Prabir Mukhopadhyay holds a BSc Honors Degree in Physiology and an MSc Degree in Physiology with specialization in Ergonomics and Work Physiology, both from the Calcutta University, India. He holds a PhD in Industrial Ergonomics from the University of Limerick, Ireland. Prabir started working with noted ergonomist Prof R.N. Sen at the Calcutta University both for his Master's thesis and later on a project sponsored by the Ministry of Environment and Forests, Government of India. It was during this time Prabir developed a keen interest in the subject and wanted to build his career in ergonomics. He joined the National Institute of Design, Ahmedabad, India, as an Ergonomist for one of the projects for the Indian Railways. There he was mentored by Dr S. Ghosal, the project lead. He then joined the same institute as a faculty in ergonomics. During his tenure at Ahmedabad, he worked on many consultancy projects related to ergonomics. Some of his clients there included the Indian Railways, Self Employed Women's Association, and the United Nations Industrial Development Organization.

After working there for two years, Prabir left for the University of Limerick, Ireland, on a European Union–Funded Project under the supervision of Prof T. J. Gallwey. He completed his PhD in Industrial Ergonomics at the same university and decided to return to India to apply his acquired knowledge. He joined the National Institute of Design, Post-Graduate Campus at Gandhinagar, India, as a faculty in ergonomics. There he headed the Software and User Interface Design discipline. He also completed a research project funded by the Ford Foundation–National Institute of Design on ergonomics design intervention in the craft sectors at Jaipur in Rajasthan, India. Simultaneously he started teaching ergonomics across different design disciplines at other campuses of the institute like Ahmedabad and Bangalore as well.

After working there for around five years, Prabir joined his present institute as an Assistant Professor in Design. He was later promoted to an Associate Professor and later became the Discipline Head. He teaches practices and researches in different areas of ergonomics and its application in design. He has authored the book *Ergonomics for the Layman: Application in Design*, published by CRC Press in 2019. He is a bachelor, and his hobbies include watching action movies, listening to Indian and Western music, traveling, and cooking. He may be reached at prabirdr@gmail.com.

Chapter 1

Introducing ergonomics to the budding designers

OVERVIEW

This chapter introduces the subject of ergonomics to the readers from the viewpoint of its application in various facets of design. It gives a broad overview of how to apply the principles, what all to be kept in mind, the steps to be followed, and the documentation of the entire process. This chapter prepares the base for the subsequent chapters in this book.

1.1 INTRODUCTION

Ergonomics, as we all know, is a multidisciplinary subject that deals with the interplay of man with product/machine/artificial elements and the environment for enhanced productivity/output/pleasure, etc. In this pursuit, man or the user is always the most important element and hence cannot be ignored. Ergonomics ensures that the product, space, communication, services, or the process that we design are reliable and are designed in tandem with human dimensions, capacity, limitations, etc.

1.1.1 Application of ergonomics in design

Ergonomics and design are "blood brothers" (Figure 1.1). Thus ergonomics plays a very important role in the design process and also when it comes to the usage of the design. We design for "humans", and thus in "any" design you need to consider humans; else your design might fail in the market. Ergonomics in design can be applied in the case of products (ranging from handheld products to bigger products like cars and aircraft), space (built space and moving space-like vehicles), and communication (ranging from static communication like signage to dynamic communication like digital interfaces). This is where the issues of identification of ergonomic issues pertaining to human dimensions, strength data along with how humans think and behave become very important in designing.

DOI: 10.1201/9781003302933-1

Figure 1.1 Ergonomics and design as blood brothers.

1.1.2 Ergonomics in tangible and intangible design

A design can have tangible as well as intangible elements. For example, the design of a hacksaw handle does not involve only material, form, and human hand dimensions (which are tangible and quantifiable) but also intangible elements like the duration one can work without fatigue, which position one should work to get the maximum benefit from the design, etc.

1.1.3 Where exactly in the design process ergonomics could be applied

Ergonomics should always be brought in the very beginning when the design is being conceived and then be applied all along the design process. It's futile to bring in ergonomics at the end of the design process. This is because at this stage the application of ergonomic principles would not be feasible in many cases, and if at all it's feasible its application would be very expensive and might not be economically viable for the company.

1.1.4 Areas of ergonomic application in product, space, communication, and services

Ergonomics can be applied anywhere, wherever humans are involved in the design, directly, indirectly, closely, or remotely. Even ergonomics can be applied in tangible and intangible design (process) as well.

1.1.5 The thrust areas to be kept in mind while applying ergonomic principles in design

As users are at the sharp end of any man–machine-environment system (refer to the book *Ergonomics for the Layman: Its Application in Design* by the same author), designers need to put themselves at the feet of the users to get to know the ergonomic issues related to the product, space, or communication and decide where exactly ergonomics could be applied.

Before starting these exercises, you need to focus on:

1. Human variation and that's where ergonomics helps you to design for a varied population
2. Identification of the human-product/space/communication touch-points and making them robust
3. Considering different contexts of use of the users and the context under which the human body undergoes different types of changes and ergonomics has to help in this
4. Cultural ergonomics aspects of the population, e.g., working or using a product in a particular way
5. Clothing and accessories worn
6. Environmental conditions
7. Who exactly are your targeting users: male, female, both, children, challenged, users with extraordinary structural features like very tall, very obese, etc.
8. The time that you have at hand to identify/solve the ergonomics issues in design
9. Resources available with you

1.1.6 Documentation

Any project on ergonomics and its application in design should culminate in proper documentation of the entire process in a manner that it can be uploaded to the design portfolio for academicians and also for the professionals so that they are able to refer to the work done later on. To do this, the reader is advised to go through the book *Ergonomics for the Layman* by the same author, which would create the theoretical grounding for doing these projects related to the application of ergonomics in design.

Any documentation on ergonomics and its application in design should follow some order like introduction, project brief, objective, methodology, results, concepts, and discussion. Every document should have a set of references at the end. Make sure to include images to supplement your document.

1.1.7 Key points

i) Ergonomics and design are blood brothers
ii) Ergonomics should come at the beginning of the design process
iii) Ergonomics helps in designing for a varied population
iv) Ergonomics can be applied in tangible and intangible designs

1.1.8 Practice session

Pick up any handheld product of everyday use. List down the good and bad ergonomic issues in them. Document your findings. You may follow the following steps:

a) Take the product in your hand and use it
b) Identify the different touchpoints with your hands
c) Identify the good and bad ergonomic issues in the design
d) Suggest ergonomic design intervention in the design with a specific focus on where exactly you would like to intervene
e) Document your entire process supplementing your document with visuals

Chapter 2

Breaking the ice ...

OVERVIEW

This chapter is an extension of the previous one. In this chapter, the exercises are done a little in-depth on the ergonomic design of products. An insight is given into the dimensional aspects of the product (anthropometry) with a specific emphasis on how to perform a task analysis and identify the product ergonomic mismatch and hence take off from there for ergonomic design intervention.

2.1 IDENTIFY ERGONOMICS AND DESIGN AROUND

Take any handheld product/space or communication modality (any piece of information) around you. Ensure that you take a product, which you have used at least once in your lifetime. Look at the product carefully and try to simulate its usage if it's not possible to use it in the studio. As you do so, highlight the different issues related to ergonomics in design.

2.2 NOTE DOWN THE GOOD AND BAD ERGONOMICS IN DESIGN AROUND YOU

Once you have the product in your access, note down the good and bad ergonomic issues in them including the context in which it would be used. See and try to justify by just applying "common sense" and do not bother about whether it's an ergonomic design issue or common sense, etc. You are just breaking the ice! So don't worry.

2.3 THE WAY FORWARD THROUGH ERGONOMIC APPLICATION IN DESIGN

Once the good and bad ergonomic issues are identified, just ask yourself what is the way forward? How can I design it much better for the user so

DOI: 10.1201/9781003302933-2

that it becomes easy, fruitful, and satisfying to use it and bring in that element of "vow" in it?

2.4 TEMPLATE FOR ASSIGNMENTS FOR DATA COLLECTION, SYNTHESIS, AND ANALYSIS

The biggest challenge for us is identifying the ergonomic design issues and translating the same into an ergonomic design solution. This translation to an ergonomic solution demands an organized approach while you collect and synthesize data from the field. The template can vary from product, space, and communication, but here we present a general template as we are only "breaking the ice".

Product

Touchpoints in the body and the object
How is it used by the body or body parts?
How it could be used for any other purpose?
Are there any chances of it being misused?
Any gender-related issues?
Issues with the elderly population
Issues with the challenged

Example: Tin can for drinking cold coffee (Figures 2.1–2.9)

Figure 2.1 The coffee can.

Figure 2.2 The coffee can in the context of the refrigerator.

Figure 2.3 Wiping of the condensed water on the can surface.

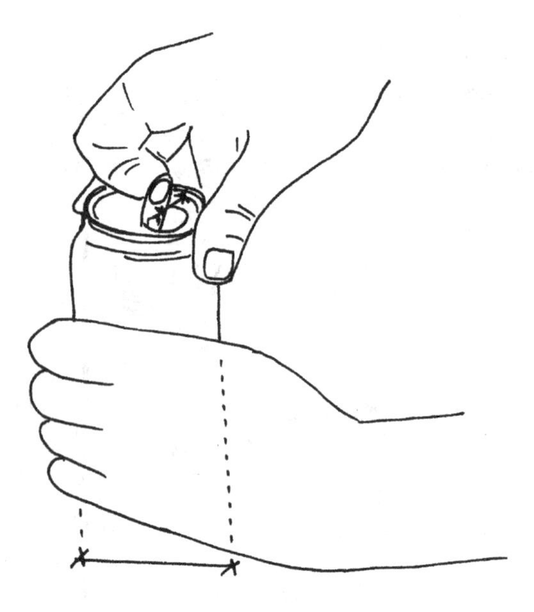

Figure 2.4 Locating the clip for opening the can.

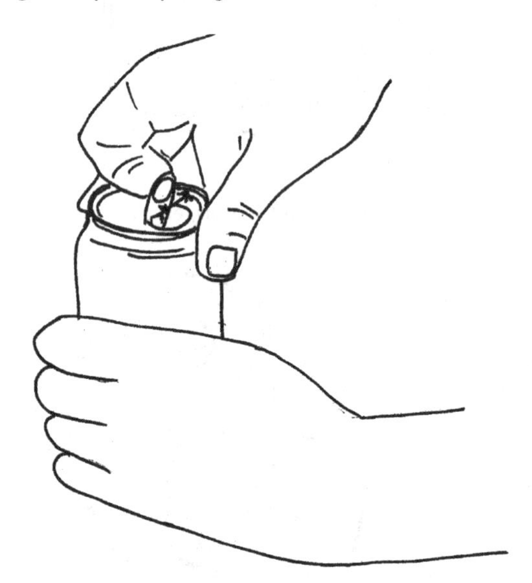

Figure 2.5 Aligning the finger with the clip and pulling it.

Step 1: pick up from the refrigerator: hand slips due to condensation
Step 2: wipe the surface with your hanky if it's too moist
Step 3: locate the clip to pull
Step 4: insert your ring finger
Step 5: Pull the clip with force
Step 6: turn the clip in another direction

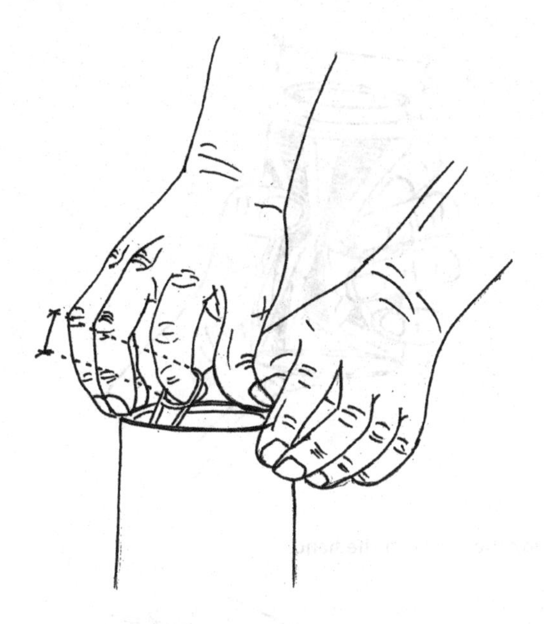

Figure 2.6 Turning the clip around.

Figure 2.7 Drinking the coffee by aligning the can with the lips.

Step 7: Take the can toward your lips
Step 8: Align the edge of the can's opening with your lips
Step 9: Drink coffee
Step 10: Crumble it and throw it in the bin

Ergonomic issues identified

Step 1: pick up from the refrigerator: *hand slips due to condensation: grip dimensions, type of grip (Figure 2.9)*

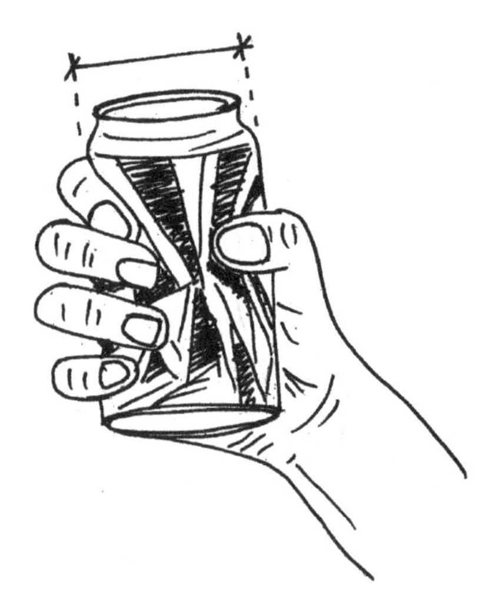

Figure 2.8 Crushing the can with the hand.

Figure 2.9 The dimension of the palm pertinent in can design.

Step 2: wipe the surface with your hanky if it's too moist: the context of use? *No one thought of it that the surface would be moist and the hands could slip*

Step 3: locate the clip to pull: *dimension of the fingertip, force required to pull, can females pull it that hard? What about the elderly population? What if one has big nails? Won't it break the nails?*

Step 4: insert your ring finger: *the diameter of the finger, fat and thin finger. Can one pull with the tip of the finger that hard?*

Step 5: pull the clip with force: *direction of force, spillage of the liquid, embarrassment in the public domain*

Step 6: turn the clip in another direction: *not as per the expectation of users! And no direction or clue that it needs to be turned in the other direction*
Step 7: take the can towards your lips: *as such no problem here*
Step 8: align the edge of the can's opening with your lips: *what about if you are in a moving vehicle? This has not been thought of! The profile of the can's opening should map the profile of lips of adults, elderly, and children as they are all different!!*
Step 9: drink coffee: *no clue how much liquid is left as you cannot see!*
Step 10: crumble it and throw it in the bin: *the material of the can should be soft enough to allow for different users to exert that force to crumble it! This kind of force would be a little difficult for females compared to males.*

So, the good and bad ergonomic issues are as follows.

Good ergonomic issues

1. Can is easy to hold in hand
2. Length of the can is convenient to hold in hand with the fluid inside
3. Easy to store in bags or refrigerators
4. Has a hygienic look

Bad ergonomic issues

1. Dimensions of the can versus dimensions of the palm not thought of (Figure 2.10)
2. No feedback from the container as to the level of fluid
3. The mechanism of opening the can is not intuitive to the target users

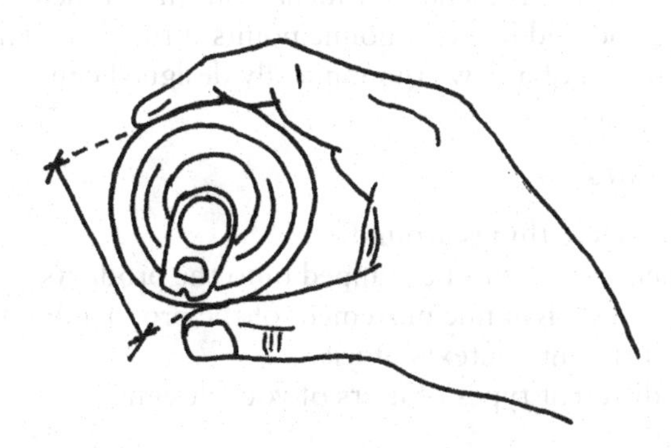

Figure 2.10 Can slipping out of the hand while trying to pick it up from the refrigerator shelf.

2.4.1 Documentation and portfolio

This would have the following steps:

A. Introduction
An icebreaking session on our first exposure to the application of ergonomics around us in any product was given. We were briefed that based on our prior understanding of science, arts, and common senses we are to identify the good and bad ergonomic issues in this product. So a readymade coffee can was selected for this project.

B. Methodology
We took the product in our hand, mimicked all the steps from holding to opening the can, drinking the coffee, and throwing the coffee can after crumbling it. For each step, the touchpoints between the human body and the product were observed carefully including the context under which it was used.

C. Results and discussions

i. The dimensions of the can both its diameter for holding and that of the clip were not in tandem with the dimensions of the palm and the fingers. This should be at par with the target users
ii. The context that the can could be moist was not thought of and the texture of the can could be changed
iii. The mechanism for opening the can should be simple and intuitive and should be easy to open for all types of users

To put this project into your portfolio:

i. Project brief: identifying good and bad ergonomic features in a readymade coffee can
ii. Through visuals represent your identification of touchpoints
iii. Point the good and bad ergonomic points through visuals
iv. Give a diagram of a new ergonomically designed can

2.4.2 Key points

i) Learn to observe things around
ii) Body dimensions are to be mapped onto the products
iii) Account for the dynamic movement of the product and the body
iv) Factor in different contexts of use
v) Consider different types of users of your design

2.5 PRACTICE SESSION

Pick up any handheld tool from the kitchen. Identify the ergonomic issues in them and give ergonomic design directions for the same, to ensure safety, productivity, and comfort of your target users. You may follow the following steps:

a) Select the product
b) Use it for the purpose it is meant for
c) Identify the touchpoints and the dynamic movements of the hand and your body
d) List down the good and bad ergonomic design issues
e) Suggest ergonomic solutions with reasons
f) Factor in the ring on the finger, use of gloves, wet and oily hand, etc.
g) Document your entire process

Chapter 3

Product ergonomics

OVERVIEW

This chapter introduces the readers to ergonomics and its application in different types of products which are used mainly by the hand and the other parts of the body as well. The products which have been used here as an example are products of daily use. How the different parts of the body play an important role in the ergonomic design of products is depicted here. The readers are introduced to the application of dynamic anthropometric dimensions by looking at the context of the usage of different products. The different anthropometric dimensions used are all "hypothetical dimensions" and have been used to show how the optimization of different dimensions is to be done. These dimensions should not be copied for designing any products. Rather students are encouraged to follow the process depicted for optimizing their product dimensions with reference to anthropometric dimensions. All dimensions unless otherwise stated are in millimeters.

3.1 ERGONOMICS IN SIMPLE PRODUCT DESIGN

By a simple product we mean any product which is "normally" or "mainly" controlled and used by the hands, but the other body parts may also play a role in its usage. The principles of ergonomics can be very well realized and felt in the usage of any simple product. The focus of this exercise should be exclusively on the following ergonomic issues:

a) Which parts of the body are involved in the product's usage?
b) What position does the body assume when using the product?
c) Do the body parts remain in a neutral position?
d) Is there an allowance for dynamic movement of different parts of the body?
e) What is the context in which the product would be used?
f) Are there any specific workwear the user uses that needs to be considered?

DOI: 10.1201/9781003302933-3

g) Are there any cultural ergonomic issues associated with the product?
h) In case the person collapses should the product dissociate from the body?
i) What Anthropometric dimensions of the body are to be considered in designing the product, both static and dynamic?

3.1.1 Hammer

Exercise
You must design or redesign a carpenter's hammer from an ergonomic perspective. You need to ensure that the user is comfortable with the hammer and gets the maximum benefit out of it (Figure 3.1).

Way forward: in steps

a) First to decide upon the length of the hammer handle and the diameter of the handle grip, the handle needs to be equated with the relevant dimensions of the hand
b) The diameter of the handle = the "grip inner diameter" (Figure 3.2)
c) The length of the handle = palm breadth + the effort arm of the hammer to be calculated for providing adequate impact (Figure 3.3)
d) Texture on the handle for adequate feedback and to enhance friction
e) Dynamicity is to be factored in for the handle length as the flesh of the palm would protrude when force is exerted on the handle
f) What different body postures would the hammer be used in? To factor this, one must visualize how the palm and wrist position changes as one uses it overhead, on the wall, etc. This aspect must be probed into
g) After you ergonomically design the product, test it out
h) The testing should be done for different types of tasks in different orientations Table 3.1 and Table 3.2

For the length of the handle, it's better to start with the higher percentile value. If the bigger hand can hold it, then the smaller hand would also be

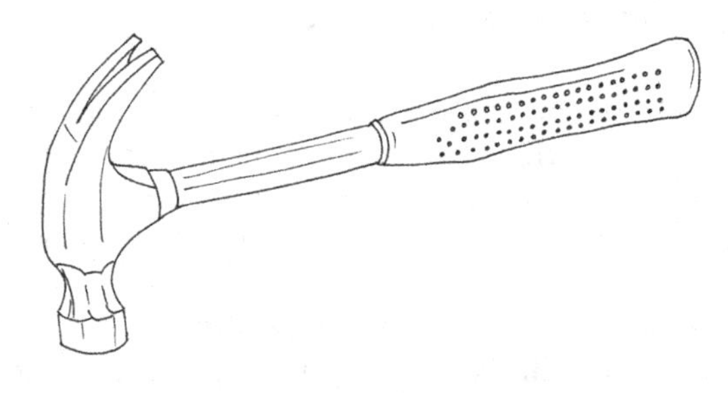

Figure 3.1 The hand-handle interface.

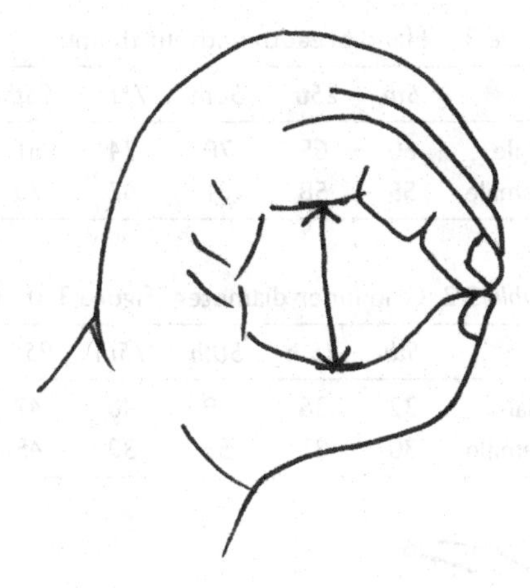

Figure 3.2 The diameter of the handle is equal to the grip inner diameter.

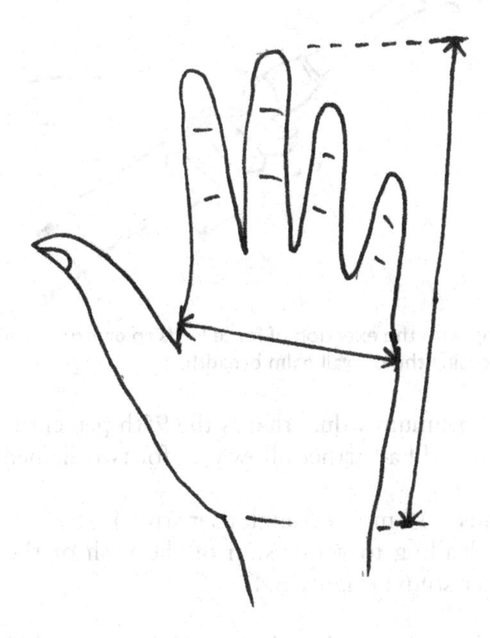

Figure 3.3 The length of the handle is equal to the palm breadth plus the effort required to use the hammer.

able to hold it (it could be a small hand holding a big handle, but still possible). On the other hand, if we start with the smallest percentile value, then for the bigger hand it might dig into the palm and the hand-handle contact surface would not be available for proper gripping.

A user study revealed that when the hand is held with a power grip, additional space is required as the flesh of the handle protrudes out.

Table 3.1 Hand breadth without thumb

	5th	*25th*	*50th*	*75th*	*95th*
Male	60	65	70	74	81
Female	55	58	61	67	70

Table 3.2 Grip inner diameter (Figure 3.2)

	5th	*25th*	*50th*	*75th*	*95th*
Male	32	36	39	40	47
Female	30	32	36	38	45

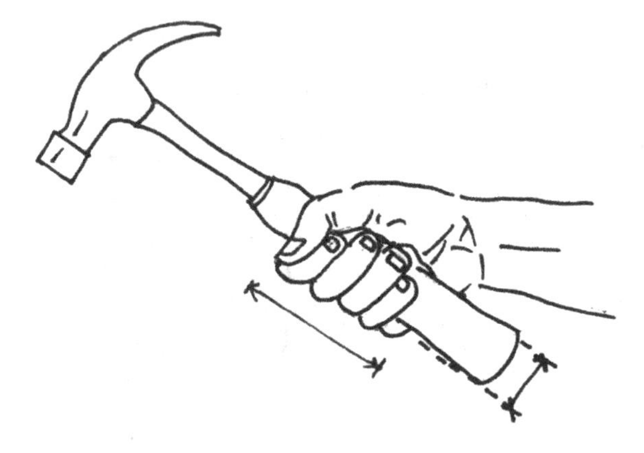

Figure 3.4 Power grip with the exertion of force leads to protrusion of flesh in the palm, thus increasing the overall palm breadth.

We go by the maximum value, that is the 95th percentile male, which is 81 mm. To this, we add a further allowance for two elements:

1. Gloved hands = 16 mm (through user study)
2. Power grip leading to protrusion of the flesh of the palm = 4 mm (through user study) (Figure 3.4)

This is an issue of access, which is the smaller hand should be able to grip it and then the bigger hand would have no problem, keeping in mind that the handle should not be too thin so that it does not dig into the palm.

Here, the lowest value is 30 mm, but to ensure a power grip and overlapping of the fingers the diameter is to be further reduced for the lower percentile values. We reduce it by 3 mm (through user study on the target users), and thus the cross section of the handle is optimized at 30 – 3 = *27 mm.*

Next: profile in tandem with palm profile and feedback

Look at your palm profile it's concave. Thus, the handle should be a little convex to "fit" the profile better. If you want to get the profile of the palm, then another way is taking synthetic mud. Make a small cylinder out of it. Size it as per the anthropometric dimensions of the palm in the previous stage. Ask around thirty different subjects of a different gender to hold it. Thus, you get different palm profiles of your subjects. You can then cut the handle from the center, which gives you an insight into the profile of the handle.

3.1.2 Door handle

Exercise

You must design the door handle which is to be aligned parallel to the door. The handle would be used to open the door. You take the door of a building for this purpose.

Way forward: in steps

a) In a door handle, there are multiple aspects. First is the placement of the handle from the floor in terms of its height. Second is the length of the handle on the door. Third is the diameter of the handle. Fourth is the clearance between the handle and the door (Figure 3.6)
b) If it's the handle of a door in a building, then it cannot be all along the length of the door because it would look ugly! And jack up the cost of the handle due to the consumption of extra material
c) Thus, handle details could be mapped like this:
Length = palm breadth
Diameter = grip inner diameter
Clearance = fist circumference (or any other anthropometric dimensions)
Height of the handle = elbow height while standing; this is how people hold and open a door. Allowance of the handle and the door is dictated by the fist circumference plus allowance for finger rings and gloved hands

Design and calculation

Start with a task analysis of how users would come and hold the handle to open the door. You will find that a user walks towards the door, body moves to the left, comes close to the door, and then grabs the handlebar to pull it open.

So now let's break the task and see the ergonomic design issues at each step:

We follow the previous example we had taken wherein we calculated the handle length and diameter as 100 and 27 mm, respectively. We can take off from this dimension as our point of reference.

a) Holding the handle

Same principle as the handle of the hammer (Figure 3.5). There the length was fixed at 100 mm (approximately). In the case of a refrigerator, a small handle would not look aesthetically pleasing. Thus, we can further increase it by say 50 mm and extend the hand as the door handle to 100 + 50 = 150 mm.

b) Clearance between the handle and the door

This is equal to the fist circumference (Figure 3.5). Or you may equate that with palm depth at the metacarpal. It's better that the optimization is towards the higher percentile value at the 95th percentile to account for gloved hands or finger rings Table 3.3.

The maximum value here is 280 mm. We can increase it to 300 mm to ensure better clearance considering different dynamic movements and the reasons mentioned above.

c) Height of the handle on the door from the floor

The best height at which a handle should be placed is at the elbow height of the user while standing. For this, select the highest of the highest value out of male and female 95th percentile and the lowest of the lowest value of

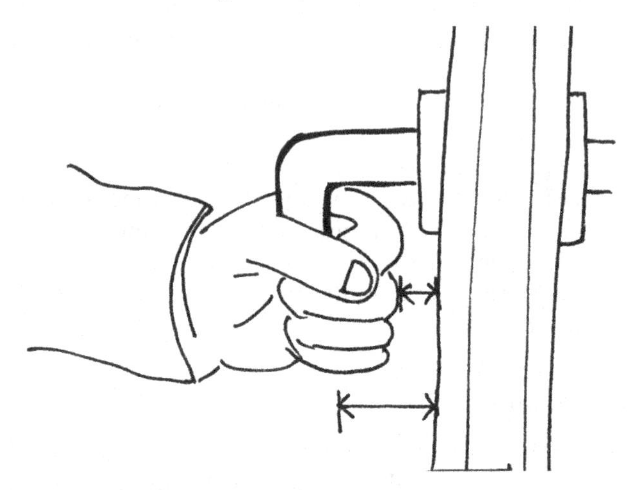

Figure 3.5 Holding the door handle with the palm.

Table 3.3 Fist circumference

	5th	*25th*	*50th*	*75th*	*95th*
Male	230	241	252	267	280
Female	198	200	210	220	236

male and female. In this case, it's 780–1103 mm and your users are within this range of values (Table 3.4).

You need to be biased towards the lower percentile (Figure 3.6), thus ensuring that the taller person's palm can go a little down while holding but a shorter person's hand should not go up while holding as it's relatively more disadvantageous. (Refer to the book "Ergonomics for the layman: Applications in design".) Thus, we start from the lowest value of 780 mm and gradually increase it upward. This is to ensure that people with higher values do not need to stoop and at the same time it's within reach of the users with lower values. A value close to 900 mm seems to be optimal as it could still be accessed by the lower percentile value given that they would be wearing shoes and the sole would add a couple of millimeters to their height from the ground. At the same time, this value would also reduce the stooping of the higher percentile values to a certain extent. This must be done through a repeated user study.

Table 3.4 Elbow height while standing

	5th	*25th*	*50th*	*75th*	*95th*
Male	840	914	1020	1066	1103
Female	780	816	829	893	1030

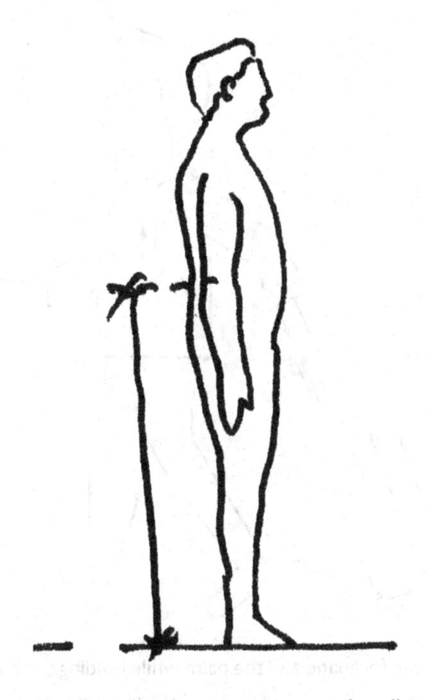

Figure 3.6 Lower percentile elbow height ensures access for all.

3.1.3 Trolley bag handle

Exercise

You must ergonomically design the handle of a trolley bag which is used by travelers. Your focus needs to be on the features of the hand-handle interface and at the same time ensure that your wrist is at neutral as you pull the trolley bag over a surface.

Way forward: in steps

a) First decide upon the dimension of the trolley bag. For your convenience, pick up any cabin baggage
b) The length of the hand-handle interface = palm breadth
c) The diameter of the hand-handle interface = grip inner diameter

Design and calculation

The trolley handle needs to be held with a firm grip, and hence there are two aspects of it. The fingers should curl around the handle and overlap. The handle length should permit the palm to fit in. The height of the trolley handle when extended should be near the waist (or trochanter) so that the user does not have to bend to hold it. Lastly the handle of the trolley should ensure that the wrist is at neutral while pulling the trolley (Figure 3.7).

Figure 3.7 Alignment of the forehand and the palm while holding the handle of the trolley bag.

If we break the above task into subtasks:

a) For holding the handle with fingers curled around the handle =grip inner diameter

The length and diameter of the handle can be fixed in the same way as seen before and optimized accordingly.

b) The profile of the handle should be convex so that it fits into the concavity of the hand

This profile must be worked with impressions of the palms or by giving a very shallow impression so that all palms can fit in easily. The addition of texture on the handle will enhance feedback to the palm and give the users a better command over the handle.

c) The height of the trolley at or near the hip level should also ensure that the wrist is not bent. If the handle is straight, the wrist might bend for users whose hip level is below the height of the trolley handle while it's extended or the hip level

The height of the trolley should be around the trochanter height (area close to your hip where you wear your trousers, the bony projection) while standing (Figure 3.8).

Here the concept is access; that is the handle should be accessible for all. If that's so, then the reference value would be 630 mm from Table 3.5 (Figure 3.8). From here, the value must be optimized in the upward direction. The reason is that though the handle should be accessible to the lower percentile value, at the same time it should not be low for the higher percentile value so that they must stoop to hold it. This accessibility of the handle should be calculated after the handle is extended as trolley bag handles are telescopic.

In Table 3.5 the lowest value is 630 mm. If we fix it at this height, then a higher percentile user at 900 mm would have to stoop and hold it, which would be difficult. This is because the difference between the highest and the lowest percentile value is 900 – 630 = 270 mm.

Thus, to bridge this gap if we optimize the height at 800 mm in that case the usage becomes easy for the higher and the lower percentile. Now the difference turns out to be 900 – 800 = 100 mm. Thus, the taller person has to stoop less and the handle is also accessible for the shorter person as well, as the shorter user can pull by keeping the hand closer to the ground compared to the taller person. Further give the entire extended handle a shallow curve so as to keep the wrist straight for users of different heights.

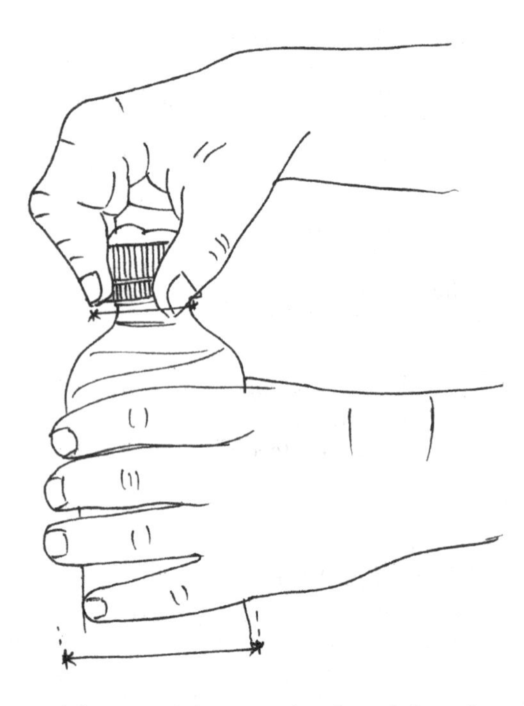

Figure 3.8 Diameter of the cap of the water bottle and the relevant anthropometric dimensions of the palm.

Table 3.5 Trochanter height while standing

	5th	*25th*	*50th*	*75th*	*95th*
Male	663	778	810	860	900
Female	630	680	700	723	819

3.1.4 Water bottle

Exercise

A water bottle is a very common product that all of us use. An ergonomic aspect of a water bottle is related to its holding, carrying, opening of the cap, and drinking out of it. The handgrip changes continuously as you hold the bottle at different stages of handling. Your focus in this exercise should be on the handgrip on the different parts of the bottle at different stages and factor in the issue that the bottle would be moist and it has to be opened in that context as well.

Way forward: in steps

a) First get your grip on a bottle inside a refrigerator the surface of which is already wet

b) The dimensions of the five fingers and the diameter of the bottle's neck should match
c) After taking out the bottle, your fingers come closer, which is a semi-precision grip
d) You open the cap of the bottle in the anti-clockwise direction, and you again require a semi-precision grip
e) Your left hand holds the bottle at the wider part, which is now equal to the grip with the fingers curled around the bottle

Ergonomic issues
To ensure that the bottle does not slip due to the moisture cover, the surface of the bottle needs to be textured. Textures apart from providing better friction also provide better tactile feedback for better grip (refer to the book by the author, *Ergonomics for the layman: Applications in Design*).

The dimension of the bottle's neck = semi-precision grip dimensions (with the fingers closed).

The take-off percentile value here is the lowest, like that of the handle diameter shown before. Gradually move from the lowest value and optimize towards the highest value and freeze at the point where your design is just comfortable for the lower percentile value and satisfies the entire spectrum of the population.

For the body of the bottle where the hand next holds, it should be close to the grip diameter (inner). Here also the take-off percentile value should be the lower percentile gradually moving towards the higher percentile in a manner to ensure that you stop at that point where the lowest percentile value cannot hold with a maximum extension of the fingers. This has been explained in the previous sections.

For the diameter of the cap, the profile might be worked upon (Figure 3.8). Giving concavity to the cap not only ensures it fit between the convex fingers and the concave profile but also aids in the application of adequate torque for opening the cap for breaking the seal for the first time.

3.1.5 Garden pruner

Exercise
A garden pruner is a simple tool, which requires grip, force, and at times torque. Thus, the profile of the tool handle is very important and should map the geometry of the palm to ensure a proper power grip over the tool with fingers curled around and, if possible, should overlap a little when closed. Your focus in this exercise should be on the power grip and the dynamicity of the tool when it opens, how far you allow it to open so that closing is possible (Figure 3.9).

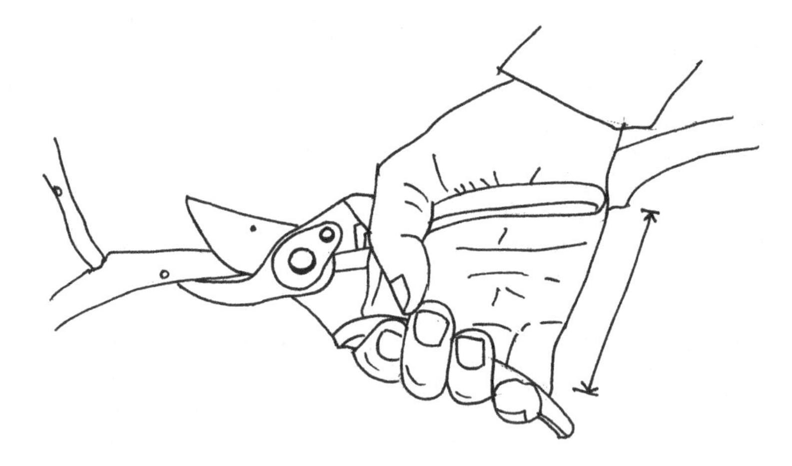

Figure 3.9 Alignment of the palm with the handle of the garden pruner considering the dynamicity in movement.

Way forward: in steps

a) The contour of the tool should be a little convex to fit the palm
b) While closed, it should be equal to the grip inner diameter
c) Use texture on the tool handle for proper feedback to the palm and to ensure proper friction between the palm and the tool handle
d) Ensure when the tool closes, the fingers do not hit one another

Ergonomic issues
The tool handle should be a little angled at 78° to the horizontal to ensure a neutral wrist position. Factor in a few millimeters (let's say 1 mm) of allowance for the usage of the tool with gloved hands. For the dimensions of the tool, you need to consider that this demands a power grip. Thus, the same principle for power grip has to be followed as before.

The length and the diameter of the handle can be worked out as explained before. Incorporate textures on the handles to ensure proper feedback to the palm and to ensure better command over the tool. Note that as the action is dynamic, it might not be possible to always keep the wrist at neutral, which is okay.

3.2 BIGGER PRODUCT

3.2.1 Refrigerator

Exercise
You are familiar with the refrigerator, which is an important component of every household today. Ergonomic issues in the refrigerator include the

height, width, and depth of the same. Apart from these, other issues include the exact height at which the handle needs to be placed along with its exact diameter and clearance from the body of the refrigerator. Your focus in this exercise should be exclusively on the above points. Keep in mind some technical feasibility in the product like the placement of the compressor, the tubing behind, etc. the position of which cannot be changed (Figure 3.10).

Way forward: in steps

a) The volume of the refrigerator is dependent on technical issues. Thus, here you do not have much role to play
b) The height of the product should be in tandem with the stature of the users (Figure 3.10 and Table 3.6)

Figure 3.10 Refrigerator and the user interface.

Table 3.6 Stature

	5th	*25th*	*50th*	*75th*	*95th*
Male	1430	1487	1537	1578	1677
Female	1331	1385	1401	1432	1521

The take-off percentile value would be lower and then gradually optimized towards the higher percentile. Keep allowance for slippers (an allowance of 2–3 cm). Thus, to the optimized value you may add 2–3 cm. Ensure that the top of the refrigerator should be accessible for the person with a shorter percentile and at the same time the person with taller stature should not have to stoop to reach for it.

c) The door handle should be at the elbow height of the user. We follow the same procedure for door handle height as discussed for doors in the previous sections

Ergonomic issues

a) Ensure that there is proper texture on the handle for adequate tactile feedback while holding the handle
b) The profile of the handle should be convex to fit the concavity of the palm
c) As refrigerator doors are now magnetic, not much force is required to open or close them. Thus the contour of the handle could be shallow enough so as to ensure a proper fit between the palm and the handle

3.2.2 Automated teller machine (ATM)

Exercise
An automated teller machine is used for withdrawing money. There are ergonomic issues in the product like the height of the screen, the keyboard, access to the card slots, and access to the slot from where the money is dispensed. You need to incorporate the ergonomic issues in this product and make it more user-friendly.

There are three or four anatomical landmarks which need to be accounted for in the ergonomic design of the product. In elevation (side and back), there are three parts of the product that need ergonomic design calibration: keyboard height, screen height, and the height of the slots for inserting the ATM card and the slot for collecting the money (Figure 3.11).

Way forward: in steps

a) Reachability is not an issue here as the person can come closer to the machine. The form of the product should allow placement of the toes and the clearance for the knees to ensure maximum proximity to the machine. See the profile of the body from the side, trace the outline, and make sure to create negative spaces at those parts which protrude out like the toe and bent knees
b) The type of tasks needs to be identified. At the ATM, it's always a "light task" and hence the work surface height should hover in and around the elbow height (while standing) of the user

Figure 3.11 Automated teller machine and the user interface.

c) For keyboard, screen, and slot height, start with the principle of access and start optimizing it from the lowest percentile value and gradually moving towards the highest value keeping in mind that the lowest percentile is still able to access it
d) You may keep some allowance for the sole of shoes (1–2 cm) and some degree of elasticity of the limb joints (1 cm)
e) The screen height needs to be fixed keeping in mind the eye height while standing and ensuring that the center of the screen is at the eye height. In the same way as pointed, start optimizing from the lowest value
f) The slots for collecting money and inserting the card should be at the elbow height and optimized with the lowest value as the take-off point as mentioned before

For deciding the placement height of the keyboard and the screen, you need to refer to the following dimensions:

Screen height = Eye height while standing
Keyboard height = Elbow height while standing Table 3.7

First fix the height of the screen with reference to the eye height (Figure 3.11). The screen should be visible to all users. The following calculation can be done to ensure this:

Table 3.7 Eye height while standing

	5th	*25th*	*50th*	*75th*	*95th*
Male	1320	1379	1425	1468	1540
Female	1210	1265	1309	1360	1409

Placement of the screen: draw a line from the eye height parallel to the ground. From the line, draw an angle of 30° above and 45° below the horizontal. This is the visual cone. Do it for all percentile values for eye height to get the common cone of vision where the screen can be placed to ensure everyone is able to see.

Placement of the keyboard: this needs to be placed at the optimized elbow height as was done in the case of the door handle (refer to Table 3.4 for the elbow height).

Ergonomic issues

The joints have some degree of elasticity for stretching forward and upward. You may add an allowance of 1 cm to the lowest value. For a user's height, you may add 1–2 cm factoring in the sole of the shoes.

3.2.3 Ladder

Exercise

You are to ergonomically design a ladder for domestic use. The ladder should be easy and safe to use and be in tandem with the human dimensions. Keep in mind the aspects of carrying the ladder and using it. Thus, you need to factor in the surface on which one would move carrying the ladder.

Way forward: in steps

a) The place for holding the ladder should be less than the grip inner diameter of the hand. Draw a table for the same and start optimizing from the lowest value as before
b) The distance between the ladder rungs is equal to or less than half of the mid patellar height. Optimize from the lowest value
c) The width of the ladder rung should be equal to the body breadth or foot-to-foot distance. Optimize from the highest value

Ergonomic issues

As we climb up, our feet are resting on the rungs one at a time except when we stand for a longer time. Thus, the power grip on the ladder is essential to marinating a proper grip on the ladder. The width of the ladder rung should support ideally at least one-third of the foot length for comfortable standing and climbing height (Figure 3.12 and Figure 3.13).

Figure 3.12 Dimensions for the distance between the ladder steps and relevant anthropometric dimensions.

3.2.4 Water faucet

Exercise

You are to ergonomically design the water faucet in a park for people going there. Remember there are a lot of elderly people who visit parks along with other types of people. You have to design in a manner that your users can have clean drinking water, and they should be able to drink it the way they want, that is directly from the faucet by hand, with a tumbler, or in a bottle.

Way forward: in steps

a) First fix the height of the faucet from the ground so that everyone is able to access it; refer to Table 3.4 for elbow height
b) The principle to follow here is reach for all
c) Then consider how you can optimize to make the life of the users with higher elbow height a little comfortable so that they have to stoop a little less

Figure 3.13 Dimensions for ladder width and relevant anthropometric dimensions.

d) The depth of the sink should take into account the elbow height as well as the movement of the forearm while drinking water
e) As one stands near the faucet, the feet should be able to move closer to the sink

Ergonomic issues
You need to factor the dynamic movement of users while standing at the faucet. The knee has to fold, and the feet have to be placed below the sink. The faucet should be placed centrally for easy access and for ease of drinking water by hand and for filling up tumblers or water bottles.

3.2.5 Key points

i) Consider reach and clearance
ii) Factor in dimensions for clothing and accessories
iii) Try to map different parts of the product with different parts of the body
iv) Ask yourself how many ways the users can use the product
v) Dimensions of the product should be in accordance with the anthropometric dimensions of the body, both static and dynamic

3.3 PRACTICE SESSION

You have been hired by a design studio for an ergonomic intervention in the design of a movie ticket counter. You must suggest ergonomic design features of the same.

Directions

a) First fix the height of the counter for the person selling the ticket while seated
b) The person selling the ticket and the customer should have an eye-to-eye contact and be at the same level
c) Raise the seat to make the salesperson sit on a platform
d) Ensure that the elbow height of the salesperson seated and the customer standing should be at the same level
e) Keep a slit on the glass window to ensure proper communication between the customer and the salesperson
f) Document your ergonomic design process
g) Indicate the limitations of your work, which could be (but not limited to):
    a. Small sample size
    b. Permission issues from stakeholders
    c. Ethics committee did not approve certain user studies
    d. Some other ergonomic aspects, e.g., cognitive issues have not been investigated

Chapter 4

Ergonomics in space

OVERVIEW

This chapter is an introduction to a hands-on approach to the application of ergonomic principles in a space. For example, a shoe store has been taken. The readers are taken to different parts of the store. Identification of ergonomic issues is shown with examples and illustrations. Different calculations are shown with hypothetical anthropometric values. All dimensions unless other wise indicated are in millimeters.

4.1 GROUNDWORK

The ergonomics of any space starts with its proper visualization. Any space has two primary ergonomic issues, namely reachability aspects for the user and the height aspects. For getting an insight into the reachability aspects of any man-made space, plan view plays an important role. This is where you can decide in which of the zones in the workspace different elements should lie. On similar lines, the height aspects of a workspace become clear if one visualizes it in elevation from the sides, front, or the back. Then calibrating the different elements in the workspace with the different landmarks in the body becomes relatively easy for fixing the work surface height.

So, while designing any workstation from an ergonomic perspective, we need to keep in mind the ergonomic design aspects as mentioned above as well as the type of task being performed, which would determine the positioning of different elements at different locations from the body.

Some directions for space ergonomics projects

Way forward: in steps

a) Take the dimension of the space with a measuring tape
b) Take dimensions of all the elements in the space, chairs, tables, washbasins, etc.

DOI: 10.1201/9781003302933-4

c) Record different types of postures that users must assume in the space
d) Record the maximum and minimum number of people in the space at any given point in time
e) Record the type of tasks in terms of their flow, pace, sequence, etc.
f) Record the path of movement of the people and staff in the space

Calculations

a) Equate first the different design elements in the space with relevant anthropometric dimensions
b) After these ask yourself, which principle you should go for. Principle for reach or principle for clearance?
c) Dynamicity involved in usage to be added over the static dimensions
d) Allowances for clothing
e) Trace the movement pattern of each user and their group
f) Calculate through different permutations and combinations of workspace envelopes for different percentile male and female dimensions. Designing a workspace envelope, you must take span or span akimbo as a point of reference and draw a semicircle around the body with the head (back of the head) as its center
g) For fixing the height of different elements, you need to superimpose different percentile male and female mannequins and then optimize
h) Grid *boards* play a very important role (Figure 4.1) in designing any space. A grid board is essentially graph papers pasted on the four walls and the floor. After you design your space on a paper, you can test your design in actual scale on the grid board. You can use different materials like polystyrene, medium density foam, and wooden blocks to create different elements in space. You can then test your optimized design on real users and refine your design further if needed

Usage of a grid board in ergonomics
After doing the ergonomic analysis of any space and coming up with a new ergonomically designed space on pen and paper, you need to test your design. This is because your ergonomically designed space is still a concept, which needs to be tested. This testing is done on a grid board on an actual scale. It's on the grid board that you can test your concept on actual users and get insights about dynamic anthropometric considerations, which are not available in the book. In other words, the grid board helps you to take your concept a step ahead towards implementation (Figure 4.2). The best is of course to design the actual space in any material to the actual scale and test it out. This is expensive at times. Thus, a grid board is a good platform to test your concept as it's very close to reality.

Approach: you need to follow the following steps for applying ergonomic principles in space design.

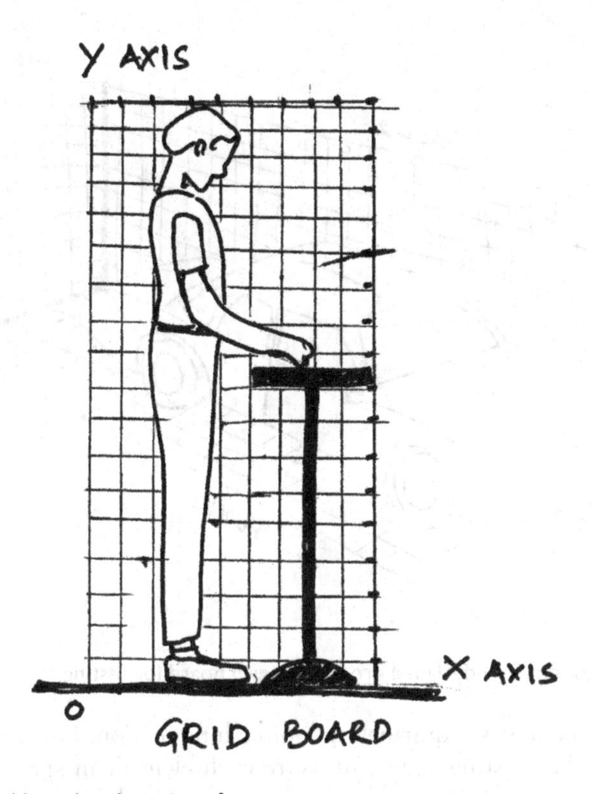

Figure 4.1 Grid board in designing of space.

A. *Analysis*
   1. Ergonomic analysis of existing space—plan and elevation drawing with all elements in space: anthropometrics
   2. Different workstations in space if any and how they are connected (link analysis—anthropometrics)
   3. Flow of people—staffs and customers in space, identify bottlenecks: anthropometrics
   4. Individual workstation—ergonomic analysis: anthropometrics

B. *Concept generation* with a focus on ergonomically designed space with efficiency in movement and comfort

4.2 SHOE STORE

Exercise

You must analyze an existing shoe store. Do a detailed ergonomic analysis, which should include the space at large and each individual element in the space. Find out the ergonomic issues in the space in terms of both good and

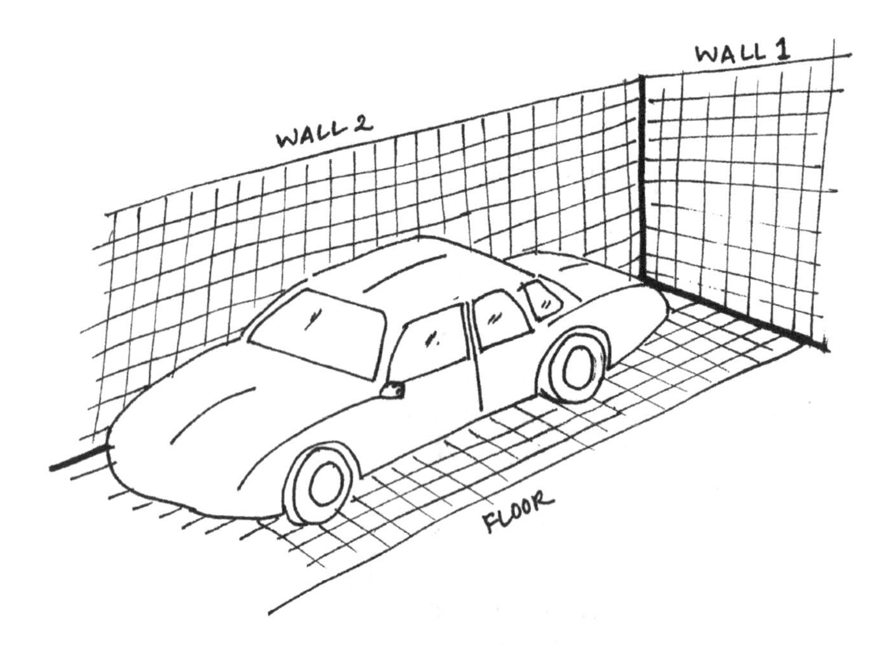

Figure 4.2 Ergonomically designed product on grid board for testing.

bad ergonomic issues requiring ergonomic intervention. For this, you need to measure the existing space, measure each element in space, then tally them with relevant anthropometric dimensions, both static and dynamic, and then redesign it from an ergonomic perspective. Remember space visualization plays a very important role. You need to visit the space at different times of the day and the week for getting an insight into the crowd flow and formation in the space.

Approach

First you need to have an overview of the entire space as to how big it is and what are the different elements in the space. The exact dimension of the store is needed so that ergonomic analysis and design directions for the same could be suggested.

Details

Area of the store: *10.7 × 4.3 m (46 sq m)*
Number of staffs: 8
Number of floors: 3
Ground floor: showroom
Second and third floors: storage

Figure 4.3 shows a glance of the showroom. It has showcases with shoes on the two sides with rows of chairs placed back-to-back at the center of

Figure 4.3 Inside the showroom, the reference space for analysis.

the space. Figure 4.4 represents the plan view of the same with different elements in space. This is the way any space should be represented for the ease of further ergonomic analysis. Figure 4.5a and b is a plan view of the storage area on the second and the third floors, respectively. The shaded are the shelves on which the shoes are stored according to their size.

4.2.1 Layout

The layout of the space gives an insight into the area, elements in space, and the exact placement of different elements in place. This paves the way for further ergonomic analysis.

As customers are the prime focus in such spaces, the seating system of the customers and the staffs was analyzed in detail.

Figure 4.6a b c d represents the customers' seating arrangement on the ground floor. It's essentially a sociopetal seating system, where customers sit back-to-back, which is good in public domain when people are unknown to one another. The dimension of this seating system was seat pan area 450 × 450 mm, seat height from the floor was 425 mm, and backrest height was 800 mm from the seat pan.

Figure 4.7a b shows the seating system from a different view and angle. Figure 4.8a–d shows the different variety of seating systems used in the shop by the staffs for different purposes like taking rest as well as for helping the customers try out different footwear.

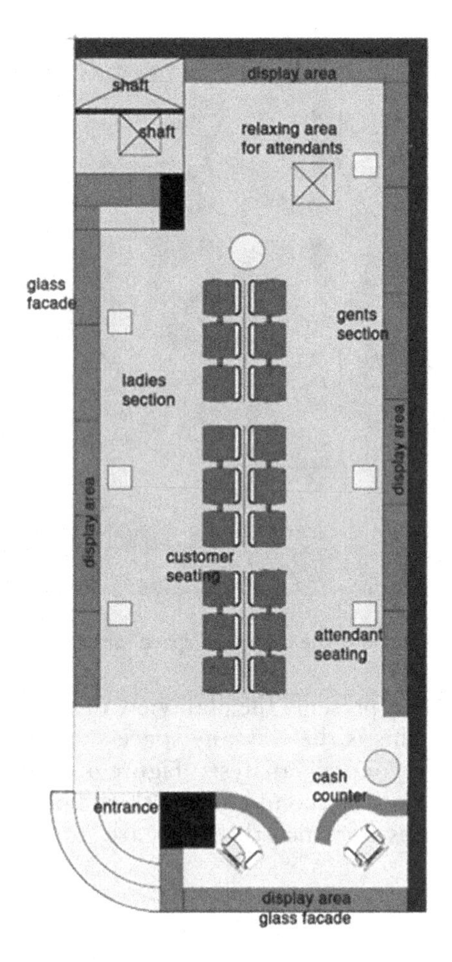

Figure 4.4 Plan view of the showroom.

The seating system of the staffs (also called attendants) was analyzed from an ergonomic perspective with a specific emphasis on the design of the same in tandem with the anthropometric dimensions of the users, customers, and staffs.

For the seats made from a transparent material (Figure 4.8a–d), the dimensions were as follows:

Seat area: 270 mm × 250 mm
Seat height: 250 mm

For the wooden seats shown (Figure 4.8e), the dimensions were as follows:

Seat area: 270 mm × 250 mm
Seat height: 250 mm

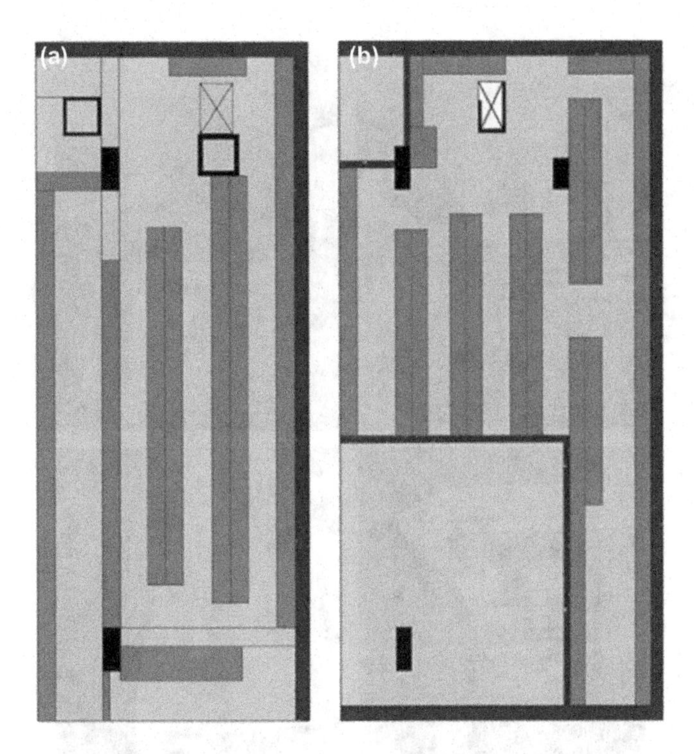

Figure 4.5 (a) Plan view of the storage area on the second floor. (b) Plan view of the storage area on the third floor.

4.2.2 Flow of people in place

The flow of people in space gives an insight into the dynamic considerations in the space and gives the designers insights into how much allowance should be provided in the space. Based on this, the customer's flow pattern was mapped at different times of the day and different days of the week and an average of this pattern is taken. Figure 4.9 a and b depicts the movement pattern of customers inside the shop. This gives an insight into the different bottleneck areas where ergonomic design intervention is required in terms of space enhancement or modification of the elements in space or rearranging the different elements.

Light line indicates the entrance and dark line the exit from the space. The photograph on the left traces the movement of a customer who enters the space, looks around, and leaves. On the right is shown a customer entering the store and trying out some shoes. There are two counters at the entry/exit. The space lacks any separate entry or exits.

Two worst-case scenarios were also studied (Figure 4.10a b). This gave an insight into the level of mismatch in the space during a festive season when the store is crowded. This is necessary to study, as the space that is designed should be able to accommodate extra crowds during peak season as well.

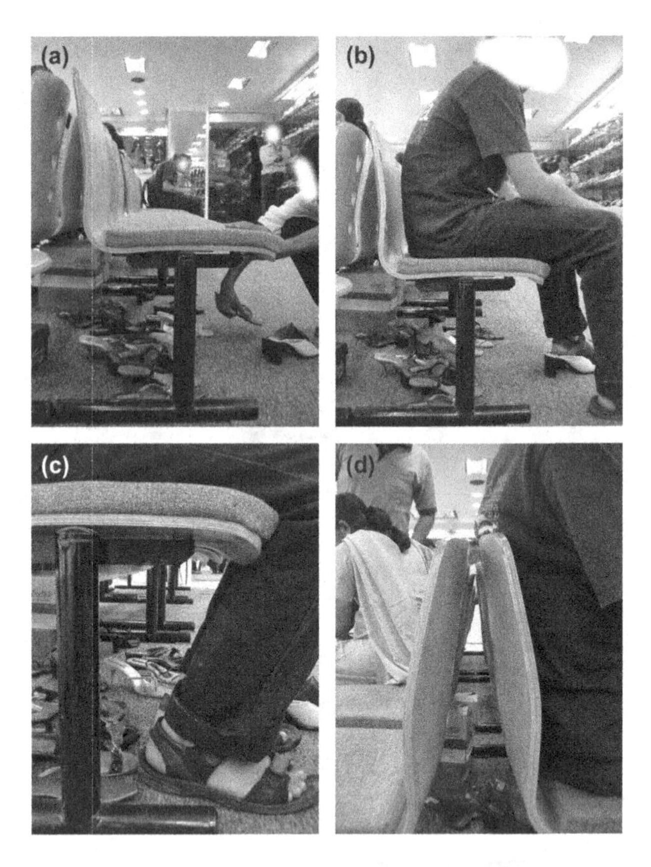

Figure 4.6 (a) Customer seating system. (b) Customer seating system with the user showing the point of contact. (c) Position of the user's feet while seated. (d) Position of the user's back and buttocks while seated.

In Figure 4.10a the light line indicates the path available for the customers to walk along the store as all the seats are full and customers are being attended for trying their shoes. Figure 4.10b indicates the zone of bottleneck indicated by a shaded circle where if the customer is attended by the staff for trying out a new shoe, and few customers stand near the cash counter, there is hardly any space left for movement. These are all depictions of different ergonomic issues in space after a user study was done in that space.

Figure 4.11 depicts the elevation of the showroom giving an insight into the dimensions of the racks especially their heights with respect to the users. This again gives an insight into the aspects of ergonomic and anthropometric dimensions necessary to optimize these designs and thus humanize them.

Figure 4.12a b c depicts a typical area of the store where customers try different types of shoes when the store is empty. It is located at one end of

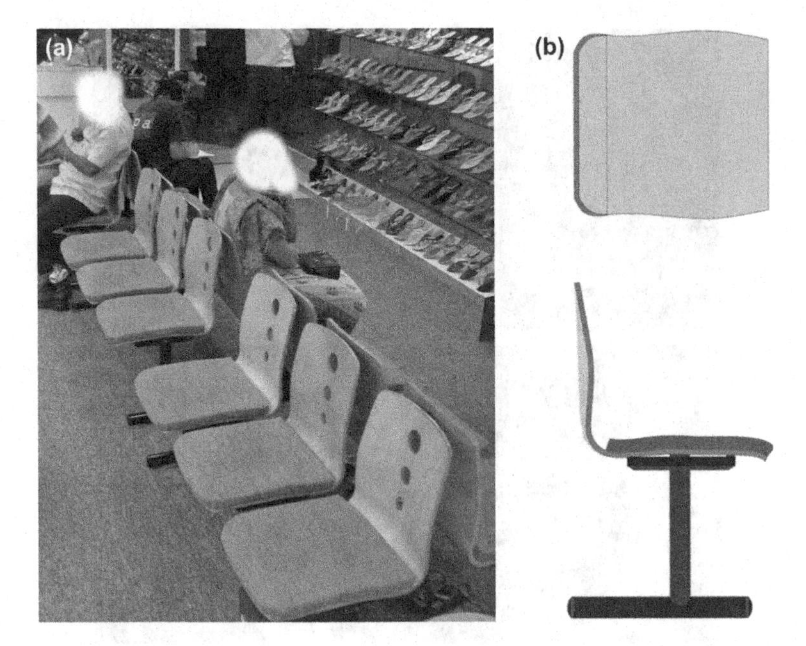

Figure 4.7 (a) Customer seating system at an angle depicting its profile and the number of seats. (b) Close up and detail of an individual seat depicting the different parts.

the store. When the store is crowded, then customers take an available seat and try different shoes with the help of the staffs.

Now that a broad overview of the store is presented in terms of the space, flow of people, different work zones, bottleneck in space, etc., a zone-wise anthropometric and ergonomic analysis is done with an eye to optimize the space and facilitate the easy movement of traffic within the store.

4.2.3 Fitting workstation

This is the dedicated place where customers try different shoes with the help of the staffs (Figure 4.13a b). The posture is shown in elevation. It shows that the customer is seated on a high seat and the staff is seated on a low seat for helping the customer in trying out different shoes. The most important issue here is to calculate how much space the customer and staff together would require in this task, which is very dynamic. This is necessary because this space behind the staff facilitates easy flow of users and can lead to bottleneck in the movement, thus creating chaos within the store.

The existing dimension of the store in terms of the width of the area distance from the back rest of customers' seat to the rack behind the staffs' stool is 1370 mm. For optimizing this space, two anthropometric dimensions were taken.

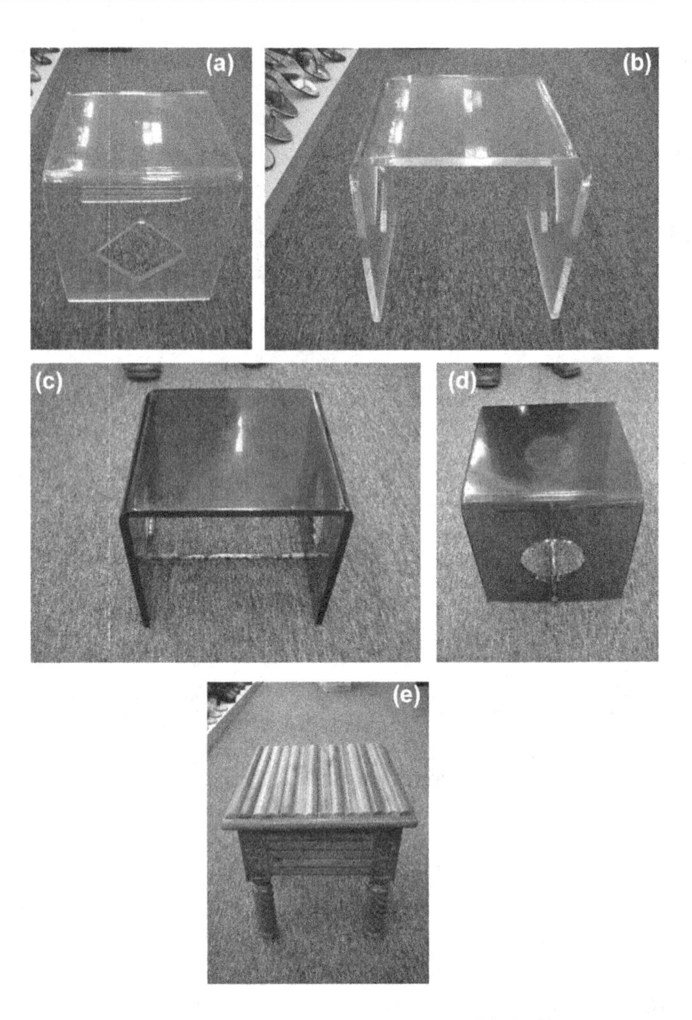

Figure 4.8 (a) Staff seating on a transparent acrylic stool. (b) Staff seating on a transparent acrylic stool from a different angle to show the space beneath the seat pan. (c) Staff seating on a dark colored transparent stool. (d) Staff seating on a dark colored transparent stool from another angle. (e) Staff seating on a wooden stool.

The buttock to extended leg length (Table 4.1) (Figure 4.14 a and b) is taken as the point of reference because this is an issue with clearance in the space. As per the table, the maximum dimension is for 95th percentile male, which comes out to be 1020 mm. This is the way a customer would extend her/his feet to the staff who would fit different shoes. Adding to this the buttock leg length of the staff (Table 4.2) (Figure 4.14a and b) would ensure an optimum workspace requirement.

So, the calculation then comes to 1020 + 685 = 1705 mm. This is the maximum space requirement if we must fit a customer and a staff in a

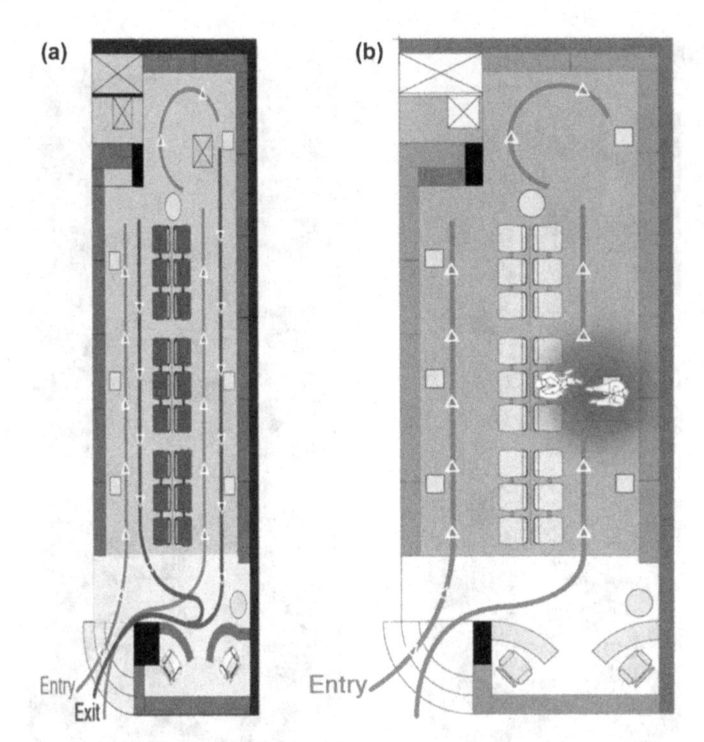

Figure 4.9 (a) Movement pattern of customers inside the space with no one seated. (b) Movement pattern of customers inside the space with a few customers seated and trying their shoes.

dynamic position trying out the shoes. The space as could be seen is calculated from the buttocks of the seated customer to the buttocks of the seated staff on the stool. This maximum dimension has been worked out considering two 95th percentile users. This is unlikely in a store-like scenario. Thus, the value could be optimized and reduced to say 1450 mm, which would accommodate users of different percentiles and would also create space without unnecessary material wastage.

Any space like the store is not static. While the task of a customer trying out new footwear with the help of a staff goes on, there are other staff members who need to move around. This is where we need to consider one person moving behind the staff while he/she helps the customer try new footwear.

To get to know the added clearance need, we may consider two anthropometric dimensions as shown in Tables 4.3 and 4.4. The maximum body breadth (Figure 4.16) is comparatively more than the maximum body depth (Figure 4.15), and hence while optimizing the clearance behind the staff for the scenario mentioned in Figure 4.15, one needs to visualize the context in both elevation and plan view. Here the maximum body depth would be a better option compared to the maximum body breadth (Figure 4.16) because body breadth is much more than the body depth.

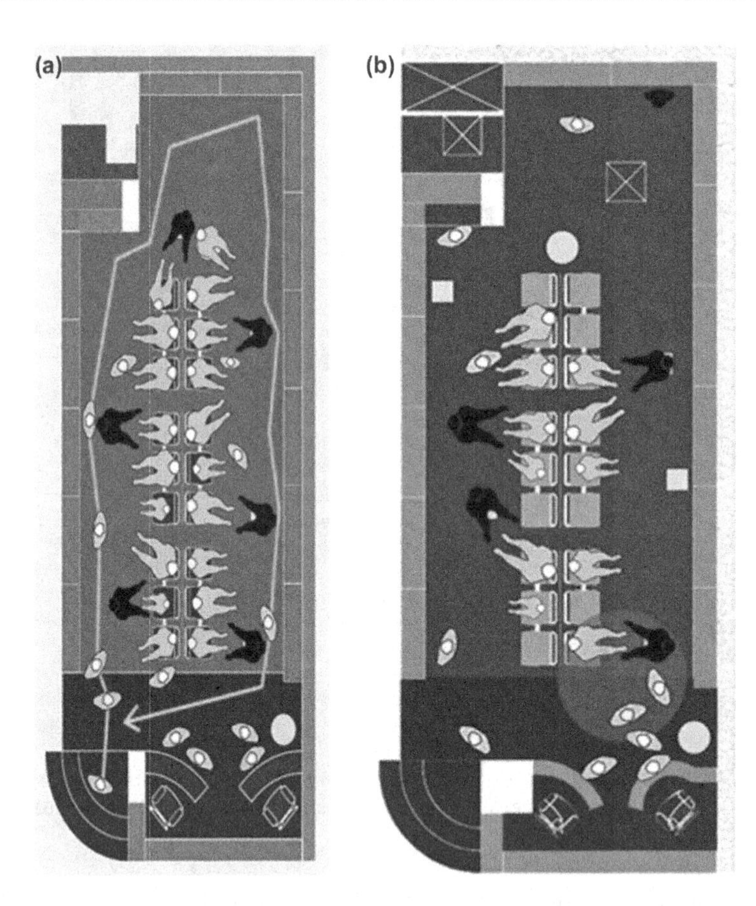

Figure 4.10 (a) Movement of crowds during a festive season in the space scenario 1. (b) Movement of crowds during festive season in the space scenario 2.

Figure 4.11 Store in elevation.

Figure 4.12 (a) Shoe trial area with the customer and the staff in a relaxed position. (b) Shoe trial area with the customer taking out his existing footwear top to try the new one. (c) Shoe trial area with the staff in action putting the footwear on the customer's feet.

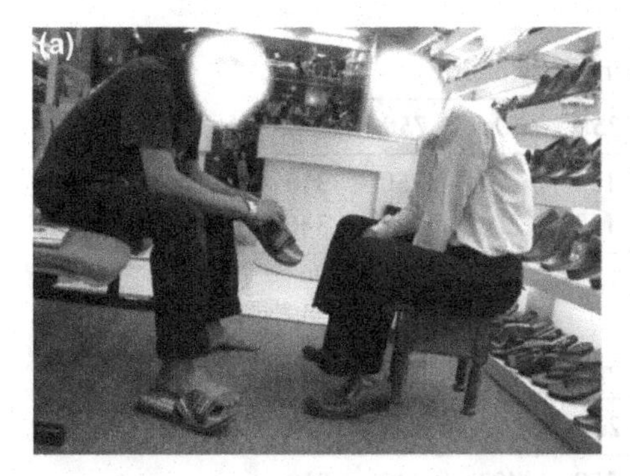

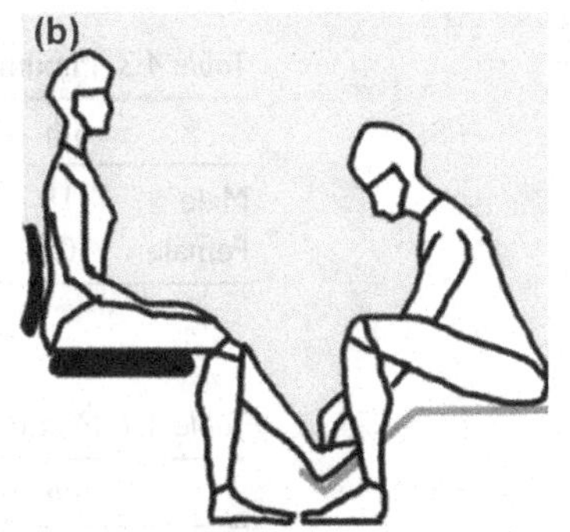

Figure 4.13 (a) Dedicated shoe trail area for specific shoes and odd sizes. (b) Shoe trial area recreated for anthropometric dimension optimization in elevation.

Table 4.1 Buttock to extended leg length (comfortable length)

	5th	*25th*	*50th*	*75th*	*95th*
Male	658	769	823	889	1020
Female	619	709	751	803	877

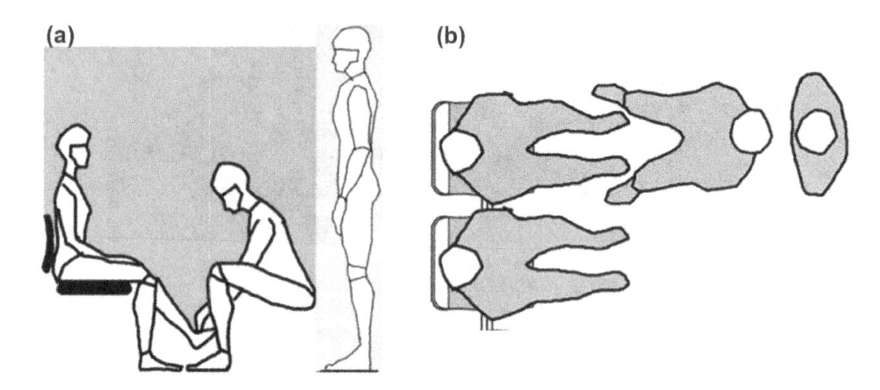

Figure 4.14 (a) Point of reference for calculating the total workspace envelope around the customer and staff. (b) Customer trying her/his new footwear with the help of the staffs (both seated) and another customer tries to pass from behind. A plan view representation of the total workspace envelope is required, which gives an indication of the aisle width required.

Table 4.2 Buttock to leg length

	5th	*25th*	*50th*	*75th*	*95th*
Male	540	555	617	677	685
Female	440	469	529	630	666

Table 4.3 Maximum body depth

	5th	*25th*	*50th*	*75th*	*95th*
Male	118	150	192	250	310
Female	107	162	220	287	340

Table 4.4 Maximum body breadth

	5th	*25th*	*50th*	*75th*	*95th*
Male	315	359	402	449	518
Female	291	323	368	418	498

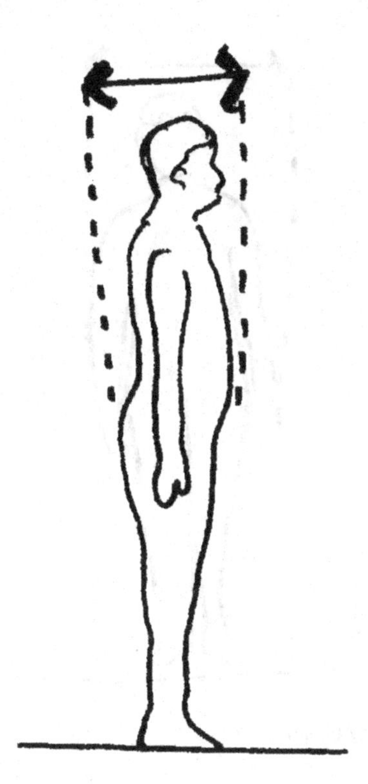

Figure 4.15 Maximum body depth.

So, we can add 310 mm (95th percentile male value): 1705 + 310 = 2015 mm. This much of allowance would allow people to move behind the salesperson with ease. Here we prefer to take the maximum values as walking involves dynamic movement of the body, and if people walk with a bag in hand, it will require some more space. The existing available dimension in the space is 1705 mm.

The next big challenge is to ensure that people with different anthropometric dimensions can walk and work in the space. One approach to doing this is to take the higher percentile value of different anthropometric dimensions related to the space. Some of these dimensions are mentioned already, and you may find out more by counting different touchpoints for people moving in such a space.

For this, we need to build different scenarios with different permutations and combinations of different percentile users and take their relevant anthropometric dimensions. From these, it becomes easy to optimize a dimension, which allows easy movement of people without any extra space. This is important because any increase in space has cost implications, as it is associated with materials and labor.

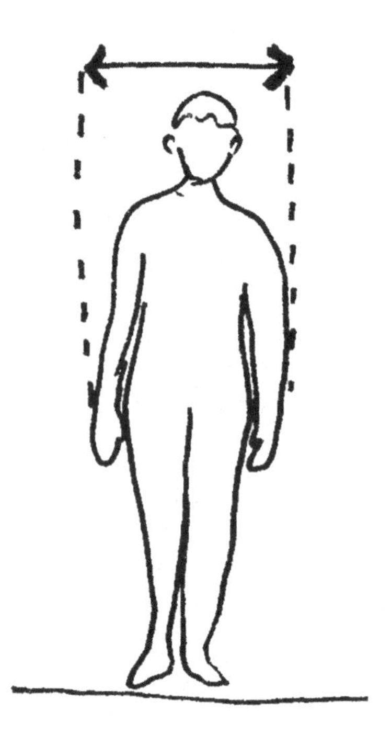

Figure 4.16 Maximum body breadth.

Figure 4.17 shows these different combinations in plan view as the plan view gives a better insight into the personal territory of users or the workspace envelope, with a specific emphasis on the movement pattern of users including dynamicity in their movement patterns. When users move, they do not move in a perfect straight line but with a bit of lateral sway. They look around the store and the racks and hence they require space for the same.

For the combinations of the four scenarios mentioned above, three anthropometric dimensions have been referred to. You may refer to a greater number of dimensions or may pick up any other relevant dimensions. The anthropometric dimensions considered are buttock to extended leg length (Table 4.1), buttock to leg length (Table 4.2), and maximum body depth (Table 4.3), respectively. The different scenarios considered were as follows (Table 4.5):

The dimension available in the store is 1370 mm. For scenario 1, there are no issues. For scenario 2 also, with a bit of adjustment there are no issues. For scenarios 3 and 4, the existing space appears to be too cramped. This justifies our increment of the workspace as suggested to 2015 mm.

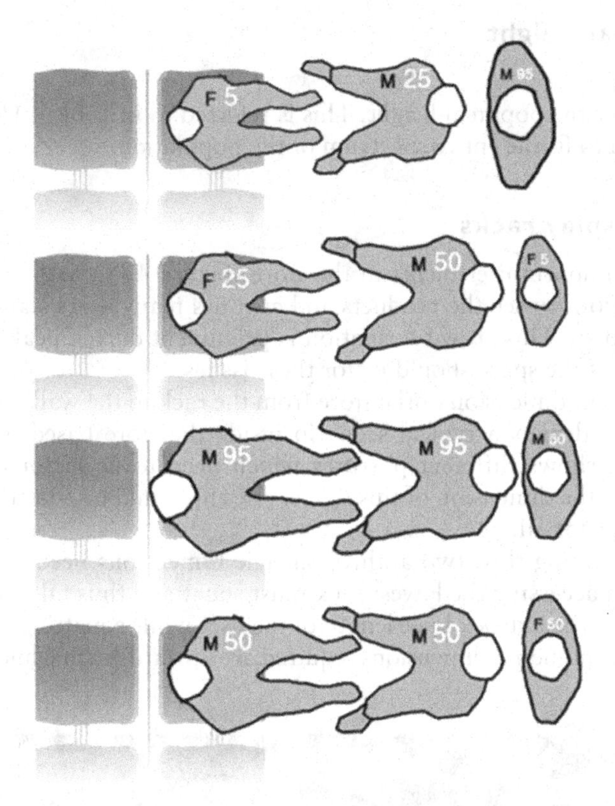

Figure 4.17 Different combinations of percentile values to optimize the space in plan view for the purpose of calculating the optimal workspace envelope. M denotes Male and F denotes Female.

Table 4.5 Different scenarios in the store while trying out new shoes

Scenario	*Buttock to extended leg length (customer)*	*Buttock to leg length (staff)*	*Maximum body depth (staff and customer)*	*Dimension required in the space (mm)*
1	Female: 5th percentile = 619	Male: 25th percentile = 555	Male 95th percentile = 118	1292
2	Female: 25th percentile = 709	Male: 50th percentile = 617	Female: 5th percentile = 107	1433
3	Male: 95th percentile = 1020	Male: 95th percentile = 685	Male: 95th percentile = 310	2015
4	Male 50th percentile = 823	Male: 50th percentile = 617	Male: 50th percentile = 192	1632

4.2.4 Seat height

Customers
We consider the popliteal height. This is referred to in Table 4.11. We fix it at 350 mm to fit the entire spectrum of the population.

4.2.5 Display racks

These are important elements in the store (Figure 4.18a and b). Users are always curious to see the products and at times many users stand near the racks to do so. This crowd formation is prominent during peak times and the design of the space should factor these issues.

The current dimension of the store from the rack to the wall is 1354 mm. Figure 4.19 depicts a typical scenario inside the store based on the task analysis. It shows different postures which need to be factored in while considering the dimension of this space; else this could be a bottleneck and create chaos for all.

For calculating this, two anthropometric dimensions need to be taken. The person accessing the lowest rack must squat and thus takes maximum space. Thus, the buttock knee length of the user while squatting is required. Next anthropometric dimensions required are upward reach while standing.

Figure 4.18 (a) Display racks with footwear for male. (b) Display racks with footwear for female.

Figure 4.19 Users at different postures at the display rack occupying different amounts of space in the common space.

Thus, we select the following anthropometric dimensions:

1. Maximum body depth while standing
2. Upward reach while standing
3. Buttock knee length while squatting

The dimensions of 1 is already mentioned. We need to now look at the dimensions of 2 and 3.

First, we need to fix the access to the uppermost rack for all. For that, the rack must be accessible to the lowest percentile, which in this case (Table 4.6) is 1420 mm (Figure 4.20). We cannot fix it at that height because it might look a little odd in the store to have such low height racks. So, we can start the optimization from 1420 mm and gradually move upwards. If we fix the height at 1500 mm, which is even less than the 25th percentile dimension for both the genders, then it is accessible for all. The difference between 1500 mm and the 5th percentile female value (lowest) is 1500 - 1420 = 80 mm. This can be managed by the 5th percentile female

Table 4.6 Upper position length

	5th	25th	50th	75th	95th
Male	1557	1688	1760	1850	1994
Female	1420	1560	1640	1752	1856

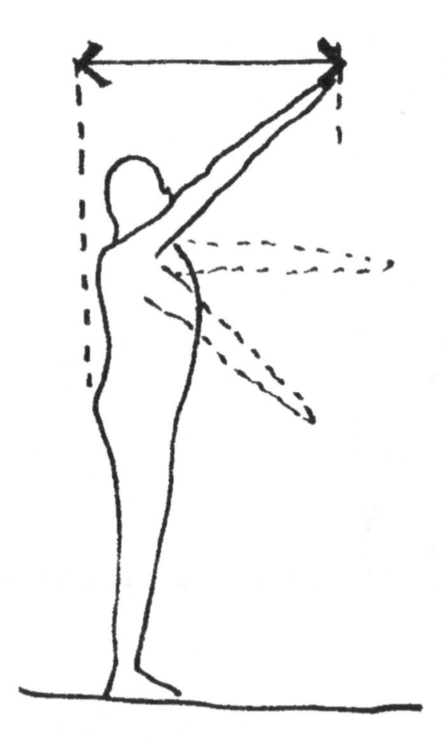

Figure 4.20 Anthropometric dimension for calculating access to the display racks while standing.

users given the fact that they wear high heel shoes and can stretch their arms a little.

If the brief of the client was not to make the racks accessible to all of them, we would have gone by the height and calculated the common cone of vision to ensure that all the racks are visible to users of different percentiles. But here we are making it accessible to all.

The next problem which pops up is that the case of the scenario mentioned in Figure 19, where users are accessing the racks in different postures and we need to provide clearance behind them. For this purpose, we refer to the dimension mentioned in Table 4.7 for buttock knee length while squatting (Figure 4.21). Here it is an issue of clearance; thus we start from the higher value. Here 470 is the highest value and we fix it there with 30 mm of additional space to allow for dynamic movement in the space. Thus, the

Table 4.7 Buttock knee length while squatting

	5th	*25th*	*50th*	*75th*	*95th*
Male	307	343	371	399	432
Female	320	359	398	430	470

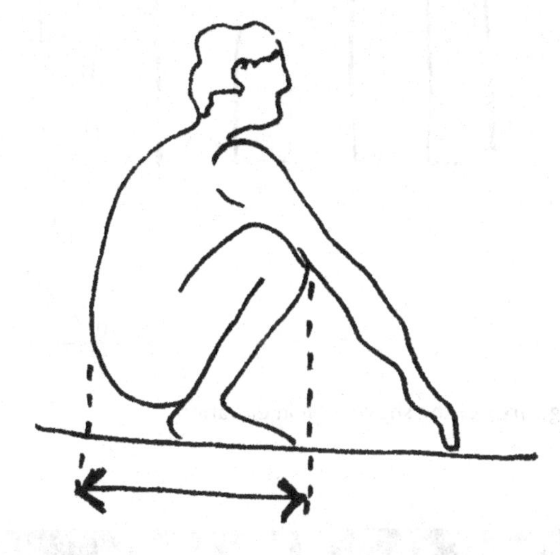

Figure 4.21 Anthropometric dimension for calculating access to the display racks while squatting.

clearance stands at 470 + 30 = 500 mm from the rack to the buttock of the person squatting.

We must further add to this value the dimension of maximum body depth as before. Refer to Table 4.3. This is again an issue of clearance and thus we take the 95th percentile male value, which is 518 mm.

Thus, the total aisle width now becomes 500 + 518 = 1018 mm, which should allow the movement of people when others are at the rack.

4.2.6 Storage area

The major ergonomic issue in the storage (Figure 4.22a) is accessing different racks by the staffs as these racks hold different types and sizes of footwear. Thus, it's very important that staffs of different dimensions should have complete access to all the racks, else the service in the store would be hampered. The two floors dedicated to storage had an inter-communicating shaft through which footwear were passed to the store on the ground floor according to the customer needs (Figures 4.23–4.26).

For fixing the height of the storage racks once again, we refer to the dimension for upward reach while standing. This is an issue of access and

Figure 4.22 Storage area of the showroom in elevation.

Figure 4.23 (a) Staff at the storage rack in the erect posture. (b) Staff trying to access the storage rack from the side. (c) Staff trying to access the storage rack in the forward bending posture.

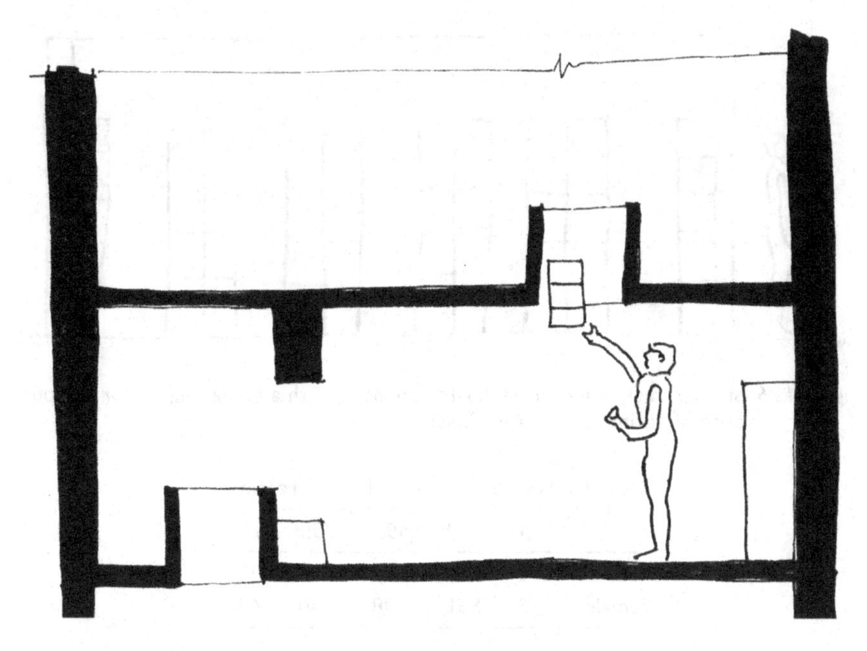

Figure 4.24 Storage racks with a shaft on the first floor.

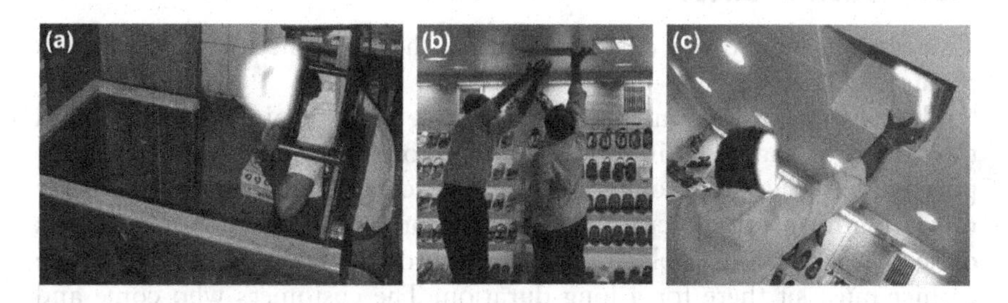

Figure 4.25 (a) Footwear transfer from one floor to another through the shaft. (b) Footwear being received through the shaft from the second floor. (c) Returning the footwear through the shaft to the other floors.

hence the height of the topmost shelf should be fixed at a little below the 5th percentile (Table 4.6) dimension for the upward reach while standing. Here it's 1420. It's better to keep it a little below this value at 1300 mm. This is because for picking up a box from the rack a portion of the forearm must go inside the depth of the rack (Table 4.8).

As this is a case of access, we take the lowest value, which is 515 mm, and fix the depth of the racks accordingly. This would ensure that people with a shorter forearm length would be able to reach the rack, grasp the box, and take it out with ease.

Figure 4.26 Storage rack in elevation illustrated along with a human figure for the purpose of optimizing the rack height.

Table 4.8 Forward reach with the forearm

	5th	*25th*	*50th*	*75th*	*95th*
Male	570	610	625	650	720
Female	515	561	588	640	690

4.2.7 Cash counter

This zone is where the customers make the payment for all purchases (Figure 4.27).

The existing cash counter was just a table and a chair. Modification in the same was suggested in tandem with ergonomic principles. It was proposed to have two steps: one for the cashier to type and write and the other raised platform for the customer to write or sign. The ergonomic design of such workstations looks at the duration of the task for the users. The cashier must sit there for a long duration. The customers who come and go need to stand for a very short duration. Thus, the workstation must be designed keeping in mind the cashier and his/her comfort.

For fixing the height of the counter for the standing customer, we must take the elbow height of the person while standing (Figure 4.28). Refer to Table 3.4.

Here fixing the table height at 900 mm would be convenient for the entire spectrum of the population, both tall and short. There is only 900 – 780 = 120 mm difference with the 5th percentile female value, which is manageable because users wear shoes and the floor has carpeting. The surface would be rarely used by customers to write anything on it.

For fixing the table height for the cashier, the elbow height of the user while seating was considered. Along with this to ensure adequate clearance between the thigh and the table top, the dimension of thigh clearance height with raised knee was considered.

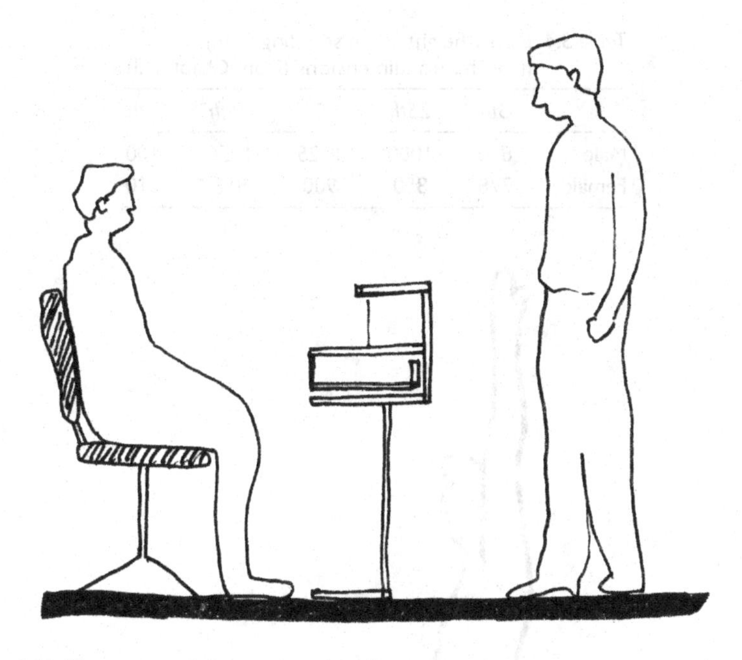

Figure 4.27 Illustration of the proposed cash counter orientation.

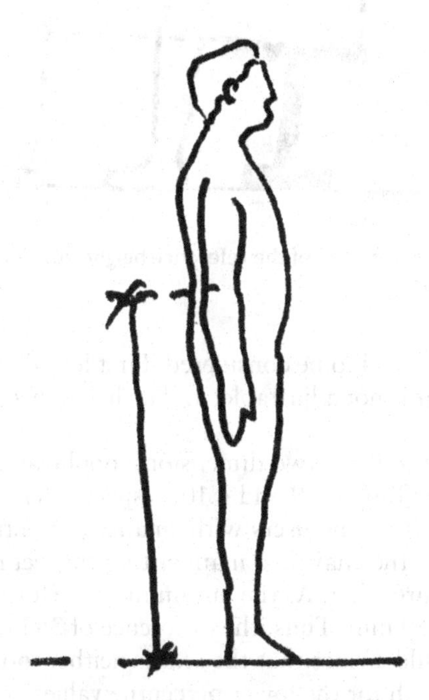

Figure 4.28 Elbow height of the customer while standing as one of the reference points for calculating the cash counter height.

Table 3.4 Elbow height while standing (mm): hypothetical dimensions (from Chapter 3)

	5th	*25th*	*50th*	*75th*	*95th*
Male	840	1000	1025	1056	1100
Female	778	850	900	918	1010

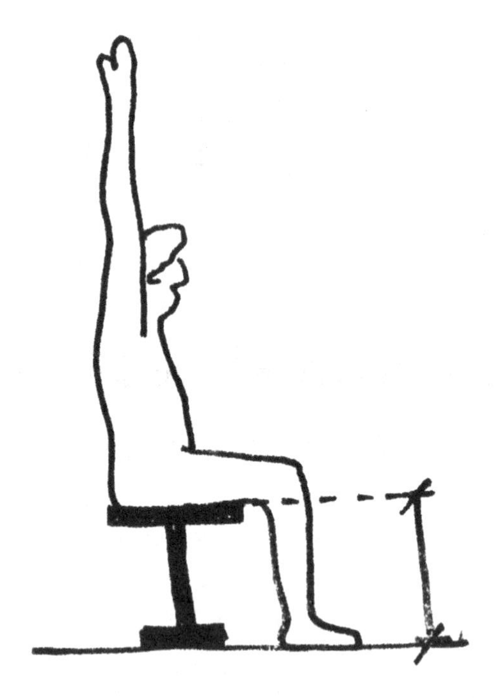

Figure 4.29 Popliteal height as one of the reference heights for calculating the seat height of the cashier.

Three parameters need to be considered. First let's fix the seat height. We imagine that the seat is not adjustable, and its height needs to be fixed. That is more tough.

For seat height, we refer to two dimensions: popliteal height (Figure 4.29) and thigh clearance (Tables 4.9 and 4.10, respectively).

Our first target is that the users with smaller percentile value should be able to sit and access the chair in a manner that the feet rest on the ground. Popliteal height ensures that. As the minimum value for the same is 263, we can optimize it to 350 mm. Thus, the difference of 263 from 350 is 87 mm. This dimension would ensure that the seat is neither too low for the higher percentile nor too high for the lower percentile values.

Once the seat height is fixed, the table height can be fixed. For fixing the table height, we take the elbow height of the person from the seat pan fixed at 350 mm. Let's say the elbow height from the seat pan at 350 mm is as follows:

Table 4.9 Popliteal height

	5th	*25th*	*50th*	*75th*	*95th*
Male	280	309	330	348	381
Female	263	285	300	320	352

Table 4.10 Elbow height while seating from the seat pan

	5th	*25th*	*50th*	*75th*	*95th*
Male	166	180	200	225	260
Female	114	162	183	198	209

Table 4.11 Thigh clearance with raised knees

	5th	*25th*	*50th*	*75th*	*95th*
Male	420	470	490	520	545
Female	401	430	450	449	500

Table height must be accessible for all especially for the lower percentile value. At the same time, it should not be too low for other percentile values as well. The lowest value is 114 mm. We can increase this value to 190 mm. Then there is a difference of 190 – 114 = 76 mm, which can be still accessed by the lower percentile considering the cushioning in the seat. This would ensure that the work surface is not too low for the higher percentile as well.

For clearance between the thigh and the tabletop, refer to Table 4.11. The table height now is 540 mm from the ground adding the seat height and the elbow height, i.e., 196 + 350 = 540 mm. This ensures clearance between the thigh and the tabletop. Thigh clearance with raised knees is the point of reference here (Figure 4.30).

4.3 ERGONOMIC ISSUES

In this selected space, there were ergonomic mismatches in terms of users and the space and the elements in the space. Mismatches were evident in terms of tasks being performed in the space. For all these, ergonomic solutions were suggested. The same approach should be taken for designing a new space for a similar task as well.

4.4 ERGONOMIC DIRECTIONS

Thus, any space could be ergonomically designed by the following:

a) If it's a new space, we need to first calculate the elements in space and their number and dimensions

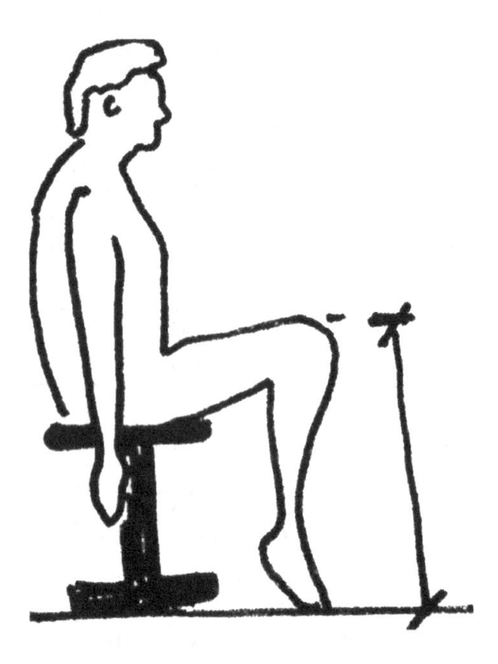

Figure 4.30 Thigh clearance with a raised knee for calculating the clearance between the thigh and the tabletop.

b) How many people are expected in the space on an average
c) Movement pattern of people
d) Postures and dynamic anthropometric issues in space
e) Task analysis of every task to be performed and the ergonomic issues in the space
f) Allowance for movement and clothing
g) So how do you move in steps:
   a. Identify the tasks to be performed
   b. Elements needed in space
   c. Postural issues
   d. Map anthropometric dimensions with space and elements in space
   e. Decide the point of optimization; clearance (higher percentile) and access (lower percentile)
   f. Your target user is male, female, or both?

4.5 KEY POINTS

i) Visualize the space in plan and elevation
ii) Consider the elements in space
iii) Map movement of people and goods in place

iv) Analyze users' posture
v) Design in tandem with anthropometric dimensions of the body, both static and dynamic

4.6 PRACTICE SESSION

Your expertise has been sought for ergonomic design intervention in an upcoming museum. The museum houses different art works of artists in two different forms: paintings and sculptures. You are to suggest an ergonomic design solution for the space and must advise the architect and give them ergonomic inputs related to the space and the display of different exhibits. You have been told to intervene in a building that is going to be double storied. The capacity of the building is such that it can accommodate 100 people at the maximum, and you need to suggest ergonomic solutions in those directions.

Directions for solutions

1. First, draw the plan and elevation of the space
2. Then decide and count the number and types of elements in space
3. Trace the movement of crowd in the place, from where you would like them to enter and from where you would like them to exit (Refer to the book "Ergonomics for the layman".)
4. Decide upon the size of the room based on 100 people at one go. Now you need to go to the plan view and calculate the primary, secondary, and tertiary zones of each percentile
5. Do various permutations and combinations of workspace envelope and decide the width between the exhibits and the length of the room
6. For deciding the exact placement of exhibits, you need to calculate the visual cone of users of different percentile values, and they optimize out of that. The best is to superimpose the eye height of different percentiles and then their visual cones in the vertical and horizontal plane. You would get the common cone of vision of the same
7. When people move, you need to allow some allowance for lateral sway of the body
8. Design barricades for guiding crowds along specific routes that you want them to go to enjoy the exhibits
9. Provide emergency exits at intervals to facilitate uniform crowd dispersion in the case of an emergency

Chapter 5

Ergonomics in moving space (transportation)

OVERVIEW

This chapter introduces the readers to ergonomic issues in moving space. The example taken here is a general compartment of a railway. The ergonomic issues in commuter transport, dimensions, safety issues, and different problems faced by the commuters are shown. Ergonomic solutions for each part of the space starting from the entry to the exit are shown with illustrations. Here also the anthropometric data used as examples are all hypothetical. The readers are to focus on the way different ergonomic issues are identified and ergonomic design solutions for the same are suggested. Use of grid boards, props, and rigs is shown in this chapter.

5.1 INTRODUCTION

A moving space is one where users work, stay, stand, sit, sleep, or even live. The only difference is that the space moves from one point to another. So, the principles of ergonomics applied in space are also applicable for a moving space as well. A moving space could be a car, bus, train, aircraft, animal-drawn cart, boat, etc. Thus, ergonomic projects related to moving space have a big challenge in that there are constraints, and one cannot go beyond a certain limit of dimensions. For example, you cannot overshoot the chassis dimensions of a truck while designing its interior. Similarly, while designing boats, you must factor in the buoyancy to ensure that the boat floats and hence cannot increase the dimensions beyond a certain limit.

5.2 SOME DIRECTIONS FOR ERGONOMICS IN MOVING SPACE-RELATED PROJECTS

As moving space is just another space with constraints, all the ergonomic principles and calculations used for space in the previous chapter applies

DOI: 10.1201/9781003302933-5

here as well. Some of the ergonomic issues which need specific emphasis in the ergonomic design of moving space are mentioned below:

a) Moving space has technical constraints and thus one cannot increase or decrease the overall space of the vehicle. Any ergonomic design intervention must be done in tandem with rearranging the elements in place, developing a new layout plan, new elements in space, etc.
b) The ground clearance, engine placement, etc., need to be strictly adhered to if a proper ergonomic design intervention is to be done
c) There needs to be more emphasis on how to "package" the occupants in the limited place and ensure their comfort, safety, and efficiency from an ergonomic perspective

One example of ergonomic design interventions in moving space is mentioned below.

5.3 ERGONOMIC DESIGN INTERVENTION IN THE GENERAL COMPARTMENT OF RAILWAYS

The general compartment of the railways in many developing countries is an opportunity area for ergonomic design intervention. It starts with the ergonomic analysis of the existing space and then suggests different ergonomic design concepts (three in this case) with different ergonomic features but with an eye to enhancing users' (customers and staffs) comfort, safety, and efficiency. These general compartments are the ones with the lowest fare and hence maximum people travel by them either to work/business or for other purposes.

Figure 5.1 indicates the layout of the train compartment in side elevation and in plan view. The side elevation shows that there are three entrances on either side of the compartment. Inside the compartment, there are seats attached to one another giving it a bench-like appearance. There are luggage racks above the seating system.

5.3.1 Movement of commuters

The movement of users in the space gives an insight into the degree of dynamicity and highlights the areas requiring ergonomic design interventions. The bottlenecks, amenities, facilities, and even the safety aspects inside the space become evident.

Figure 5.2 indicates that users enter through the two doors in each compartment. They then move to occupy a seat if there are any. In case seats are not available, they stand and try keeping their luggage on the luggage racks. If users need to go to the toilet or drink water, then their movements are along a defined path. The washroom and toilets are located at one end

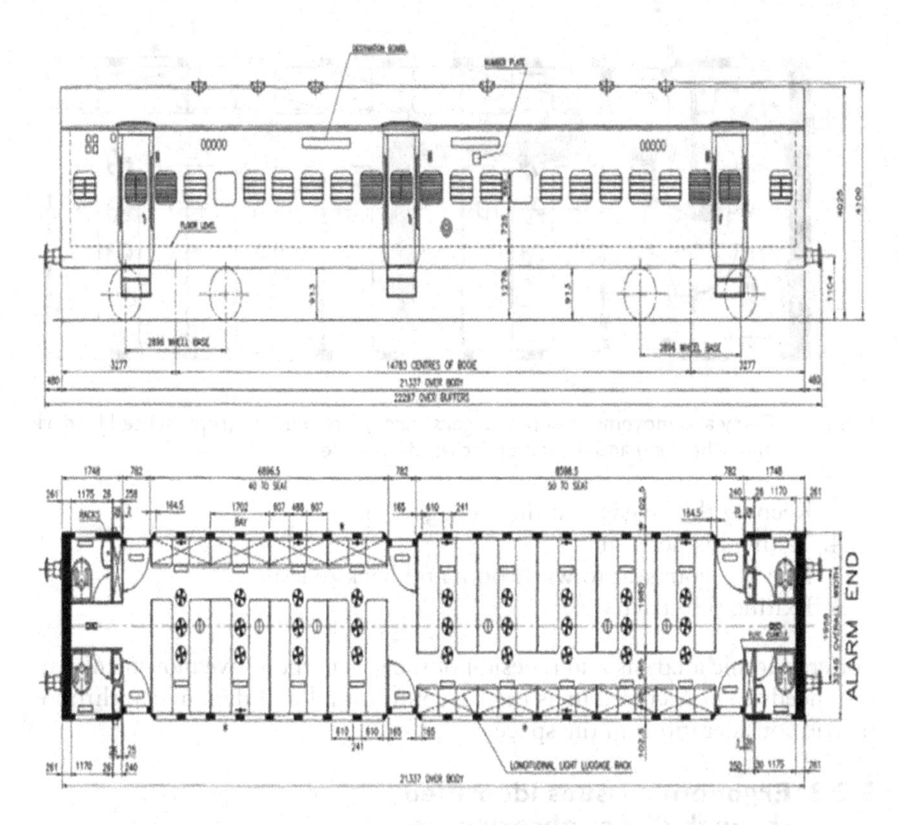

Figure 5.1 Existing layout of the general compartment studied in elevation (top) and plan (bottom).

of the compartment as shown in Figure 5.2. This movement for accessing the amenities creates a bottleneck (indicated by superimposing dark and light lines) in the space when commuters enter the space or get down from the space at different stations.

The space has a capacity of seating for 90 users. The luggage racks are meant for keeping luggage only. As these spaces are overcrowded, more than 160 users occupy their seats, and more than two users use the luggage spaces for sitting. All these were observed through direct observation and activity analysis by the research team who traveled on this train for several trips.

5.3.2 Task flow analysis

Task flow inside the compartment comprised of the following steps:

1. Entering the coach
2. Moving from the corridor to the seat

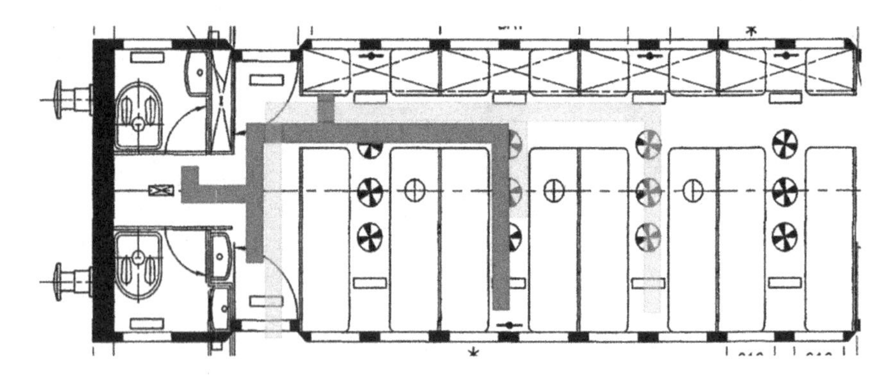

Figure 5.2 Entry and movement of passengers through two doors (represented by dark and light lines) and movement toward the toilet.

3. Keeping the luggage on the luggage rack
4. Taking his/her seat
5. Moving from seat to washroom and back to seat
6. Exiting the coach

For ergonomic analysis and redesign of the space, the above movement pattern needs to be factored in as they account for a lot of dynamic anthropometric considerations in the space.

5.3.3 Ergonomic issues identified through direct observation

Some ergonomic issues were identified in the space through direct observation and indirect observation through video and still photography. For this purpose, the cameras were fixed at different locations inside the compartment. While clicking photographs, you need to be careful to frame the problem areas in context.

The major ergonomic issues identified and as indicated in Figure 5.3a–e are as follows:

1. No proper chain handle: the emergency chain did not have any proper grip or handle. Thus, it was difficult for users to grip and pull it in the time of emergency
2. Inadequate space in the corridor with luggage: when the luggage was stacked in the corridor, due to lack of space in the luggage racks, it was often stacked in the corridor.
3. Improper emergency windows: the emergency window was smaller than the 5th percentile bideltoid dimension of the users
4. Improper placement of washbasin: the washbasin was placed near one of the entrances and the door was hinged on that side and thus could not be opened completely

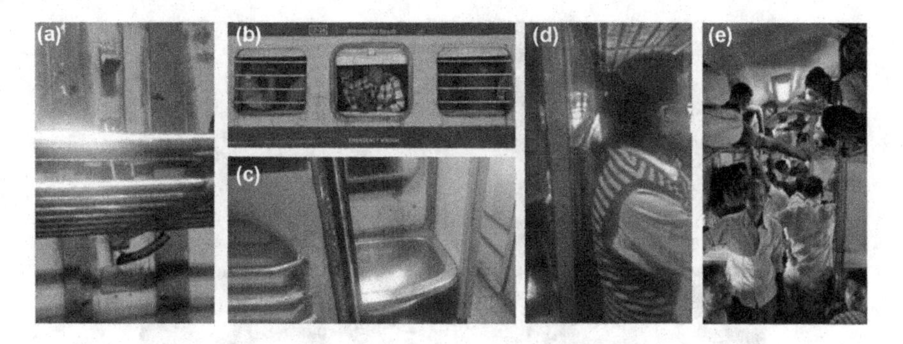

Figure 5.3 (a) Placement of the luggage rack. (b) Emergency window for exiting the compartment not in accordance with adult anthropometric dimensions. (c) Placement of the sink behind the main door of the compartment. (d) Backrest of the passenger seat is straight and not cushioned making it uncomfortable. (e) Crowded compartment showing passengers' orientation in different manners.

5. Improper seats leading to shoulder and neck pain: the seats were not designed in tandem with the anthropometric dimensions of the users and thus they were not comfortable, and users were complaining of pain in different parts of the body

Figure 5.4 indicates that the entrance to the coach is very narrow and leads to bottleneck. This is because the psychology of the users in this part of the world is that everyone tries to board and get down the train together. Thus, it creates this bottleneck. Moreover, when users must move along this entrance with luggage or kids, it becomes even more difficult because of the narrow width. Thus, with an eye to these problems an optimization for the space was suggested.

In Figure 5.5, an exercise for the optimization of the entrance and corridors was carried out. The left side was optimized for male passengers and the right side was optimized for female passengers to see where the optimized value reached. Span akimbo was taken as the reference anthropometric dimensions (Figure 5.6 and Table 5.1).

For calculating the workspace envelope of the users, the space is examined in plan view. To get an insight into the workspace envelope of users of different percentile values, first draw the span akimbo in plan view to scale. You may scale it down to 1:5 or 1:10/1:20 as convenient for you. To do this, you just have to divide the original value by 5, 10, or 20. Once you draw the span akimbo, then from the back of the head, taking half of the span akimbo (head to elbow in plan view) draw a circle around the user. Do this for different percentile values. This circle gives you an insight into the individual workspace envelopes. This is what is represented in Figure 5. The different permutations and combinations of users' envelopes for both

Figure 5.4 Crowded entrance at the entry/exit door of the compartment.

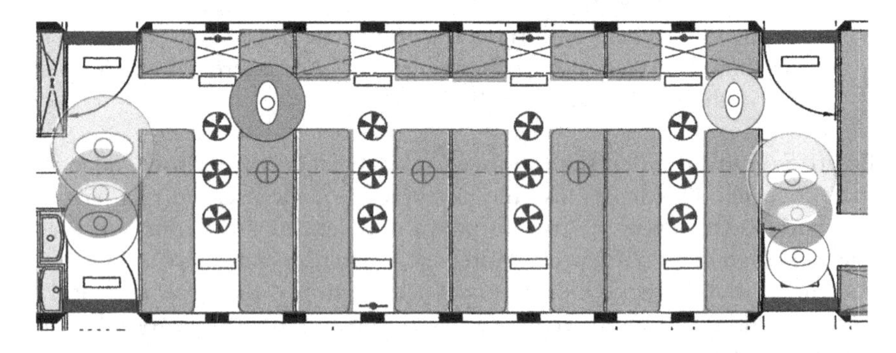

Figure 5.5 Optimization of the entrance and corridors in plan view. Different workspace envelopes are indicated by white 5th percentile, dark gray 50th percentile, and light gray 95th percentile. Left side for male and right side for female passengers.

genders help us to arrive at the optimized value for the space where everyone would be able to navigate with ease.

Since this is a space, it's an issue for clearance. That means if "fatter people" can pass through, then "thin people" would not have any problems at all. You see this mapping of different percentile workspace envelopes in the figure and get to know for which percentile values the space is cramped. Based on these, two sets of explorations were done with male and female users' dimensions on the two sides of the compartment.

Taking span akimbo as the point of reference

The existing dimension of the corridor is 782 mm.

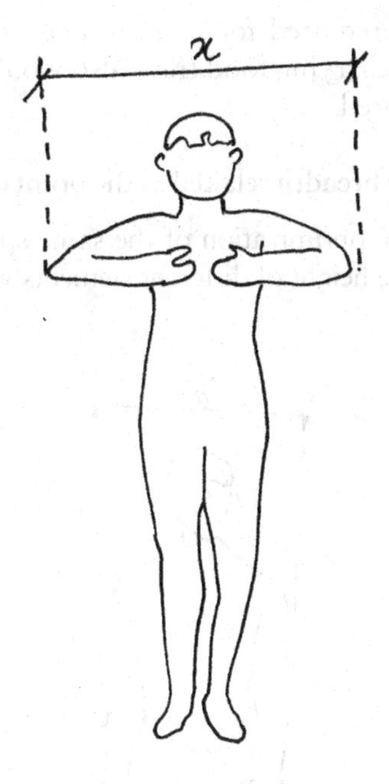

Figure 5.6 Span akimbo taken as one of the reference points for optimizing the door width and the corridors.

Table 5.1 Span akimbo

	5th	*25th*	*50th*	*75th*	*95th*
Male	665	730	760	800	888
Female	590	650	690	740	790

This is an issue of clearance, and hence we should take 888 mm as the point of reference for optimization.

Let us take two contexts:

Person carrying a duffle bag: dimension of the bag = 190 mm
Person carrying a baby in lap: dimension of the baby = 200 mm

So, these extra dimensions are to be added to the span akimbo dimension, and it comes to:

Since 200 mm is maximum, we add it to span akimbo of 95th percentile value: 888 + 200 = 1088 mm.

So, this is the space required for users to enter the space and navigate through. Even vendors carrying food trays also would be able to navigate if this dimension is proposed.

Taking maximum body breadth relaxed as the point of reference *(Figure 5.7)*

Figure 5.8 represents the optimization of the same space in elevation so that we get an insight into the height of different elements with respect to the body.

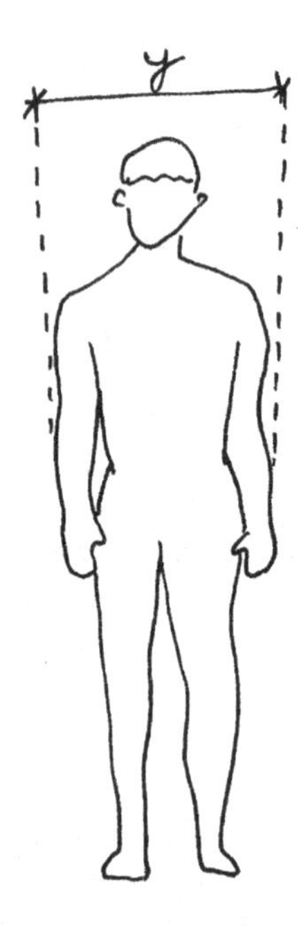

Figure 5.7 Maximum body breadth for calculating the entrance and corridor to account for the dynamic movement of passengers and vendors.

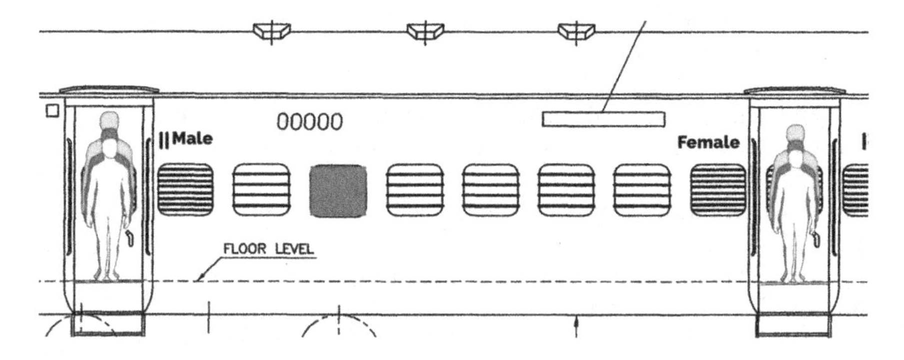

Figure 5.8 Optimization of the corridor with maximum body breadth relaxed.

If we add 200 mm with the maximum value 520 mm: 520 + 200 = 720 mm.

In Figure 5.9, the scenario demands the usage of two different anthropometric dimensions. One is the maximum body depth and the other the maximum body breadth (Tables 5.2 and 5.3). The sum of two dimensions would ensure free flow at the gate and corridor when people are standing there. Here we need to consider the maximum body depth plus the maximum body breadth of the users to ensure the free flow of users in the space.

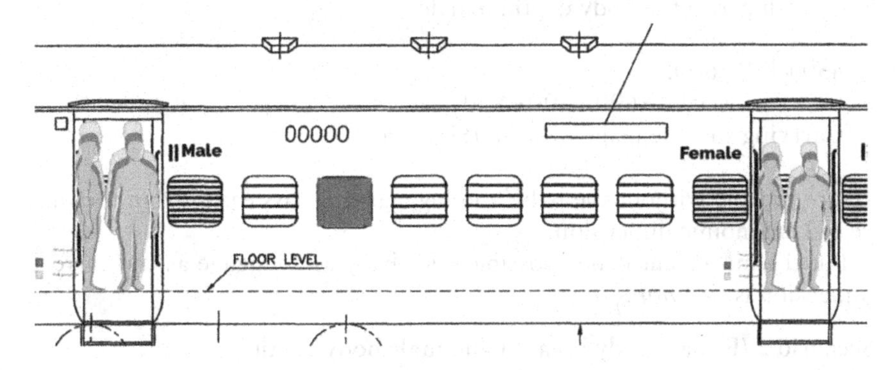

Figure 5.9 Scenario when one person is standing, and another person wants to pass through.

Table 5.2 Maximum body breadth relaxed

	5th	*25th*	*50th*	*75th*	*95th*
Male	312	358	406	450	520
Female	292	325	370	392	502

Table 5.3 Maximum body depth

	5th	*25th*	*50th*	*75th*	*95th*
Male	119	140	200	256	300
Female	109	150	210	270	340

Let's look at different permutations and combinations of different percentile users.

Scenario 1 (Male body breadth and body depth)

Figure 5.10 is a close-up view of the entrance/exit to the coach and represents its look and feel.

Details of the scenario:

- 75th percentile maximum body breadth: male
- 25th percentile body depth: female

450 + 150 = 600.
Carrying a baby = 600 + 200 = 800.
Carrying some luggage = 600 + 190 = 790.

Hence, people without the baby and luggage can pass easily from the suggested ergonomic dimension.

But the movement is not possible with baby and luggage as the current dimension is 782 *mm*.

Scenario 2 (Female body breadth and male body depth)

75th percentile maximum body breadth of female = 392 mm.

25th percentile maximum body depth of male = 140 mm.

Total space occupied = 392 + 140 = 532 mm.

If they carry a baby, then 532 + 200 = 732 mm. It's possible because the dimension is 782 mm.

And if carrying a baby is possible, then carrying a bag is also possible (532 + 190 = 722 mm).

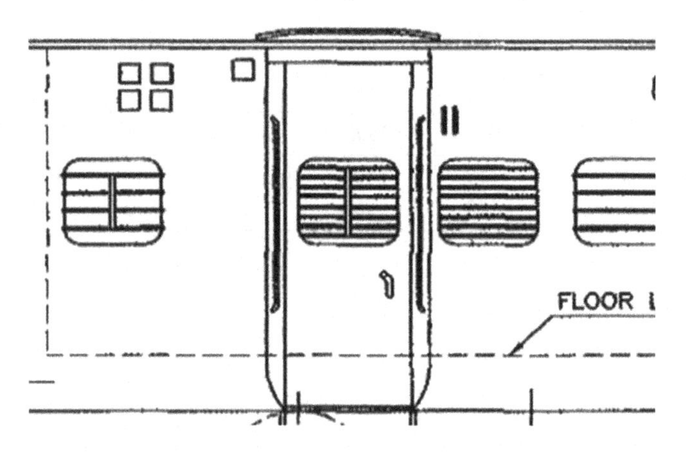

Figure 5.10 Entrance/exit of the coach.

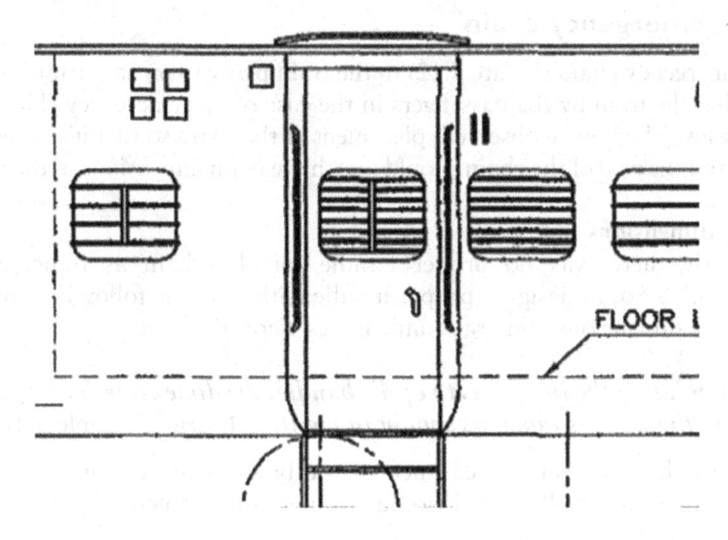

Figure 5.11 Optimized dimension of the entrance and corridor.

Scenario 3 (Female body breadth relaxed and female body depth)

- 50th percentile body breadth relaxed: female (370 mm)
- 25th percentile body depth: female (210 mm)

370 + 210 = 580.
Carrying a baby = 580 + 200 = 780.
Carrying a suitcase = 580 + 190 = 770.

Hence, the movement is possible only without the baby and luggage and when carrying the baby and luggage in body depth position.

As the current dimension is 782 mm, movement with a baby and a suitcase is possible.

In this way a greater number of scenarios could be created to see the best possible optimization that would cater to the need of the maximum spectrum of the population. Here we have shown only three scenarios. You may go for more to test out how users of different percentile dimensions would fit in the space.

There are different optimized values and thus it is presented in Table 5.11. Shown in Figure 5.11

Figure 5.11 shows the optimized dimension of the entrance and the corridor after explorations of different percentile users for the specific anthropometric dimensions. The optimized dimension now stands at 830 mm which would cater to the need of all different types of users including the dynamicity involved in all movements with luggage, trays, and kids.

5.3.4 Emergency chain

The emergency chain (Figure 5.12) in the train plays a very important role in stopping the train by the passengers in the case of any emergency. There are two issues which were observed: placement of the chain so that it's accessible to all and the grip of the chain should match the palm dimensions of the users.

Chain dimensions

Currently there was no proper handle for the chain as indicated in Figure 5.12. So, to design a proper handle with grip, the following anthropometric dimensions and ergonomic issues were considered.

For calculating the inner width of the handle, the dimension handbreadth without thumb was taken as shown in Figure 5.13 (right) (Table 5.4)

Here it's the question of clearance. If the bigger hand can go inside the handle, then the smaller hand would not have any problem.

Figure 5.12 Emergency chain located just below the luggage rack and close to the wall.

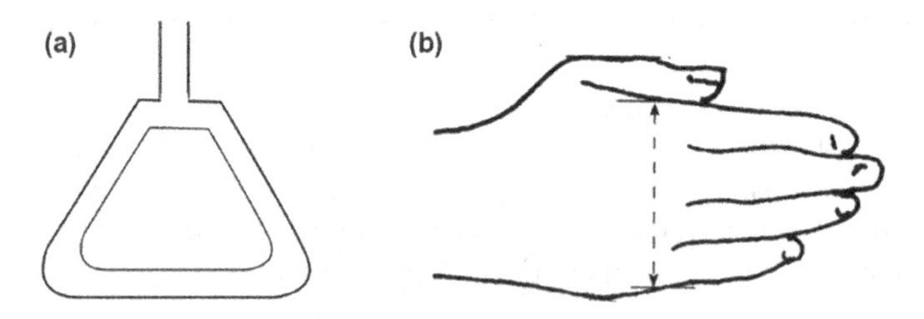

Figure 5.13 (a) Profile of the handle suggested for pulling the chain. (b) Anthropometric dimensions of the palm considered for designing the chain handle and its profile

Table 5.4 Handbreadth without thumb

	5th	*25th*	*50th*	*75th*	*95th*
Male	60	65	70	74	81
Female	55	58	61	67	70

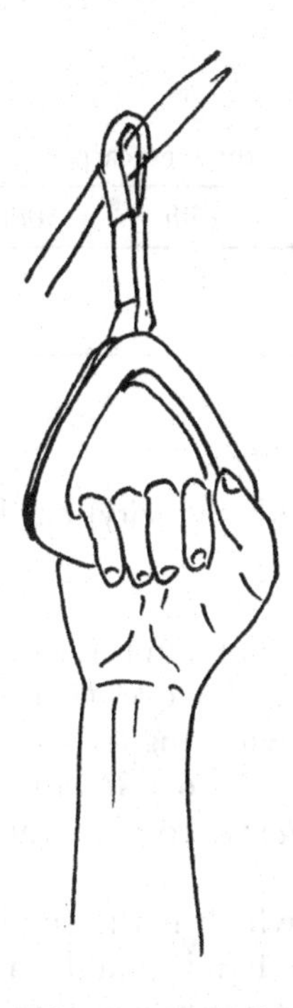

Figure 5.14 Flesh of the palm protruding out when the handle of the chain is held with a power grip. This dynamicity needs to be considered in designing the handle.

User study revealed that when the hand is held with a power grip, an additional space is required as the flesh of the hand protrudes out of the handle (Figure 5.14).

We go by the maximum value, that is 95th percentile male, which is 81 mm. To this, we add further allowances for two elements:

1. Gloved hands = 16 mm (through user study)
2. Power grip leading to protrusion of the flesh of the palm = 4 mm (through user study)

For calculating clearance of the handle (Figure 5.15), hand depth at the metacarpal was taken

This is again an issue of clearance, and we take the maximum value (Table 5.5).

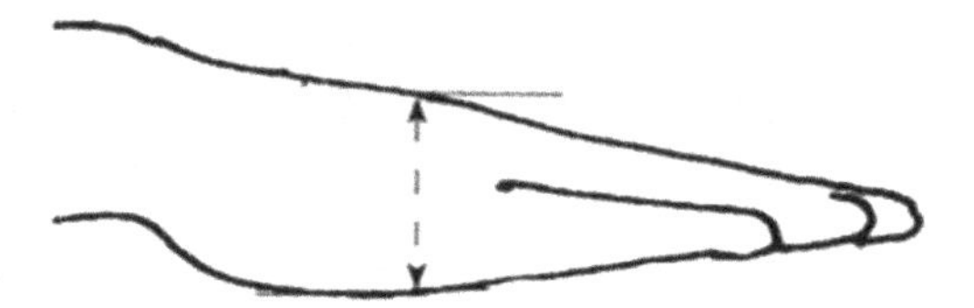

Figure 5.15 Hand depth at the metacarpal for the clearance inside the handle.

Table 5.5 Hand depth at the metacarpal

	5th	*25th*	*50th*	*75th*	*95th*
Male	8	11	14	16	20
Female	7	10	13	15	18

The maximum allowance (dynamicity) has been considered after a user study on the target users:

Clearance for max height with hands (+6)
Clearance for max height with gloves (+4)
Clearance for max height with ring (+6)
Total allowance = *20 + 6 + 4 + 6 = 36 mm*
Cross section of the handle: we consider the grip inner diameter

This is an issue of access, which is the smaller hand should be able to grip it and then the bigger hand would have no problem, keeping in mind that the handle should not be too thin so that it does not dig into the palm.

Here the lowest value is 30 mm, but to ensure a power grip and overlapping of the fingers the diameter is to be further reduced for the lower percentile values. We reduce it by 3 mm (through user study on the target users), and thus the cross section of the handle is optimized at 30 – 3 = *27 mm*.

Chain handle height

The next task is to decide at what height from the floor of the coach the chain should be fixed so that at the time of emergency everyone is able to pull it. Thus, the optimization is done by considering two available anthropometric dimensions.

Upper position height of a person standing in the erect position

The chain must be accessible for all. So, the lowest value here is 5th percentile female (Table 5.6), which is 1420 mm. We can expect the person to be wearing footwear with a sole thickness of 10 mm (Figure 5.16).

The total height turns out to be 1420 + 10 = *1430 mm*. This is when the person is standing erect and trying to access the chain, and this is the height keeping in mind the shortest percentile. The superimposed

Table 5.6 Grip inner diameter

	5th	*25th*	*50th*	*75th*	*95th*
Male	32	36	39	40	47
Female	30	32	36	38	45

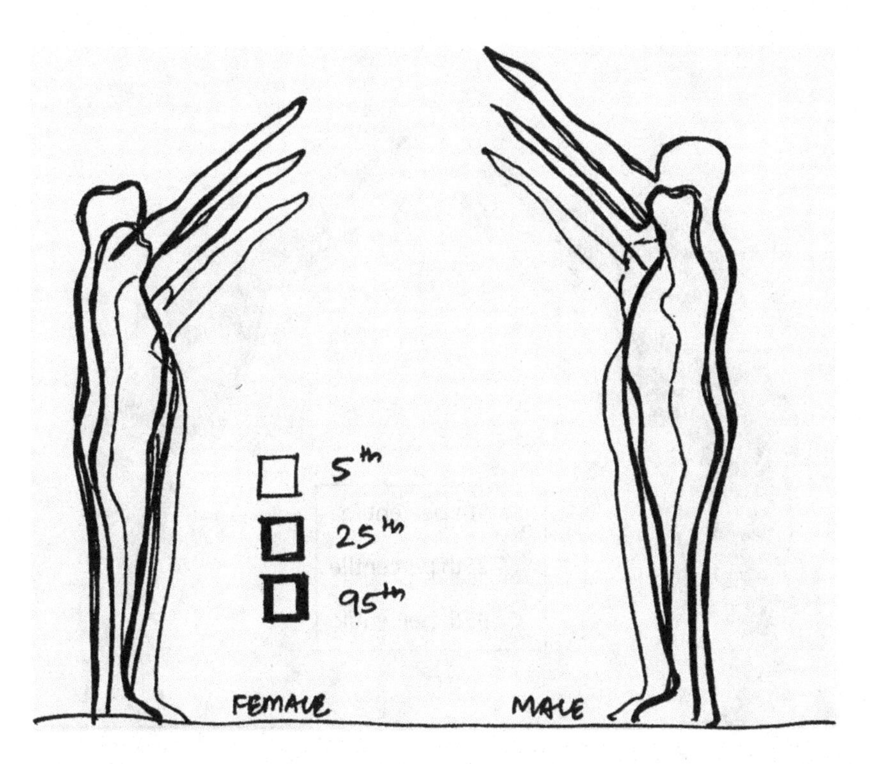

Figure 5.16 Upper position height in the erect position. Superimposed values of 5th–95th percentiles for female (left) and male (right) users.

values of different percentiles if done to the scale give you an estimate of the difference between the different percentile values for different users (Table 5.7).

Standing in the front-leaning posture

If the user must access the chain while leaning forward due to the crowd in front who are either women or children, then the dimension in Figure 5.17 is to be considered (Table 5.8).

Table 5.7 Upper position height of person in the standing erect position

	5th	*25th*	*50th*	*75th*	*95th*
Male	1560	1700	1760	1851	1972
Female	1420	1560	1662	1756	1850

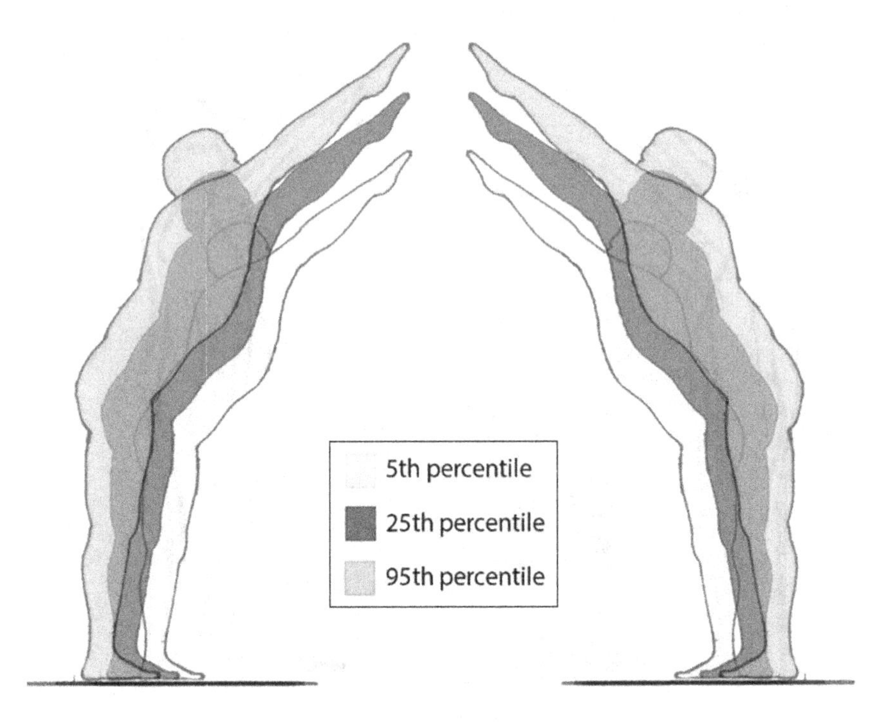

Figure 5.17 Standing in the front-leaning posture. Superimposed values of 5th–95th percentiles for female (left) and male (right) users.

Table 5.8 Upper position height of person in the front-leaning position

	5th	*25th*	*50th*	*75th*	*95th*
Male	1430	1560	1650	1740	1998
Female	1309	1458	1560	1640	1770

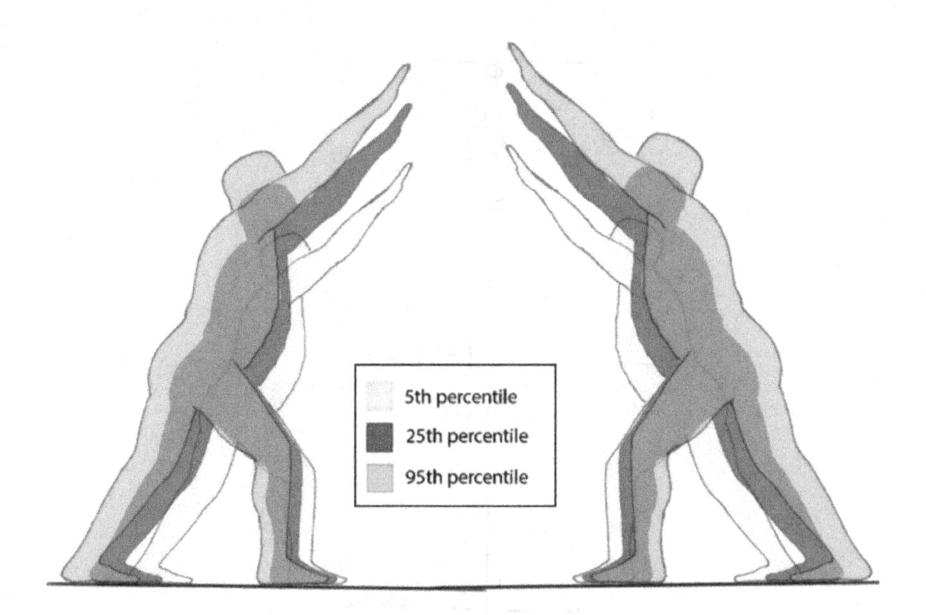

Figure 5.18 Standing with one step in the forward-leaning position.

Here the lowest value is 1309. If we add 10 mm for footwear height, it becomes 1309 + 10 = *1319 mm*. The chain should be at this height if the person must access it in this position.

Standing with one step in the forward-leaning position

This is not a recommended position as it's difficult for users to pull the chain in this posture. Here the lowest value must be taken.

Figure 5.18 represents the different percentile values for male (right) and female (left), giving an insight into the difference between different percentile values (Table 5.9).

The lowest value is 1220 mm. We add 10 mm for footwear: 1220 + 10 = *1230 mm*.

Table 5.9 Standing with one step forward-leaning position

	5th	*25th*	*50th*	*75th*	*95th*
Male	1380	1550	1640	1730	1850
Female	1220	1490	1580	1670	1790

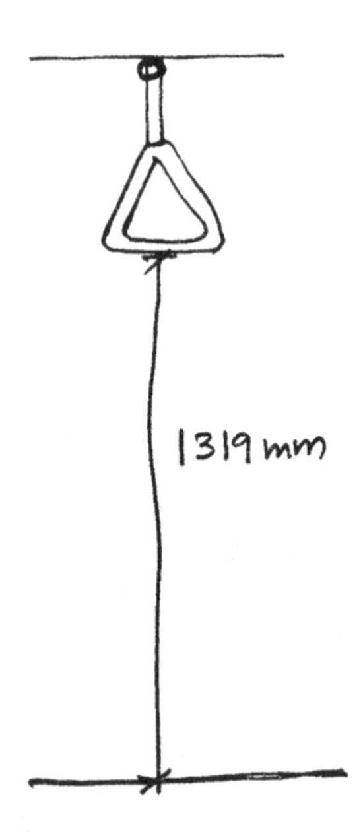

Figure 5.19 Optimized dimension of the chain.

After exploring different postures at which users stand inside the compartment, the optimized value is 1319 mm. At this height, it's accessible to all and is not too low inside the space (Figure 5.19).

5.3.5 Ladder for climbing to the upper berths/seats

One needs to take care that the spacing between the ladder rungs (Figure 5.20) should be such that it's easy to climb for those in the lower percentile range.

No data base was available to refer to the gaps between the ladder rungs. If there is, you may refer to starting with the lowest percentile as this is a question of access for the smaller percentile.

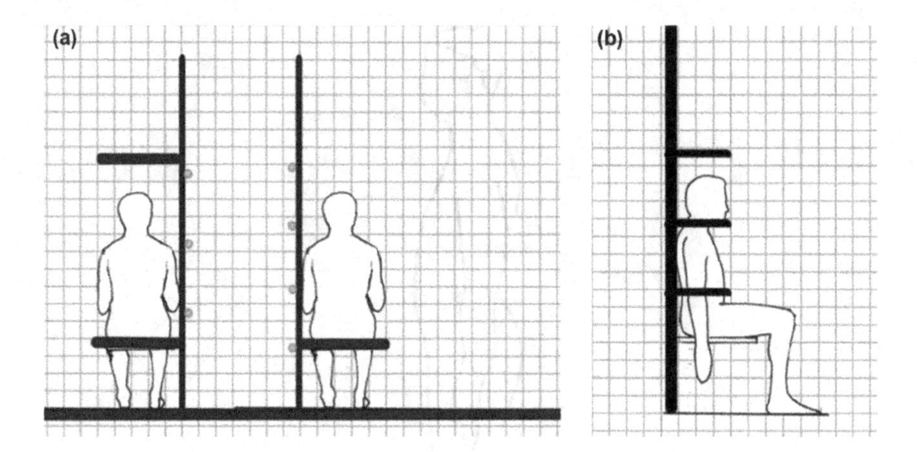

Figure 5.20 (a) Concept of the ladder for climbing the upper berth: rear view. Ladder for climbing on the upper berths (left cross section and right side view). One concept. (b) Concept of the ladder for climbing the upper berth: side elevation

Here we have taken the mid patellar height (Figure 5.21) as the point of reference and have further reduced it from that value (Table 5.10).

The lowest value is 258 mm for 5th percentile female. This dimension is also not comfortable for the lowest spectrum of the population for climbing the ladder. Thus, we reduce it to half the value. It then becomes 258/2 = *129 mm.*

So, the spacing between the ladder rungs was fixed at 129 mm to make it accessible for the entire spectrum of the population.

The diameter of the handle of the ladder remains at *27 mm* as was calculated for the chain.

5.3.6 Layout of the compartment to enhance the ease of movement and passenger comfort

After doing an ergonomic analysis, the mismatches in terms of anthropometric dimensions are identified. The task analysis has given some insights into the degree of dynamicity needed in the space and the needs and wants of the users. Keeping all these in mind the final layout of the space is now suggested. The dimensions are again hypothetical, and it's the process that's important.

5.3.6.1 Concept 1

In this concept there have been all explorations within constraints (Figure 5.22).

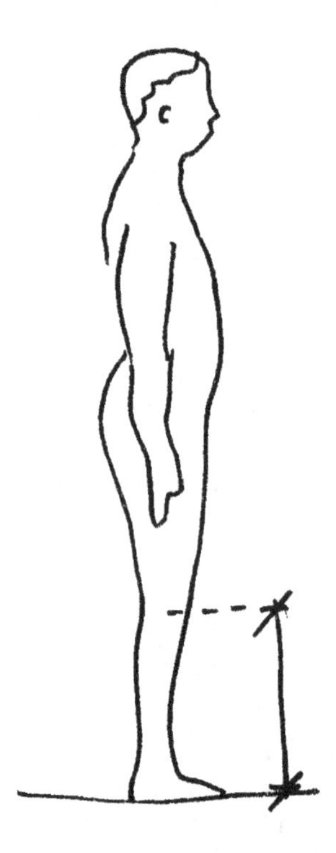

Figure 5.21 Mid patellar height as one reference point for calculating the distance between ladder steps.

Table 5.10 Mid patellar height

	5th	*25th*	*50th*	*75th*	*95th*
Male	319	330	358	370	462
Female	258	302	326	349	382

The default seat length was taken with the introduction of seat division between seats in the form of a small ridge of 80 mm width. The length of the corridor for ensuring the ease of movement was taken at 560 mm, and earlier it was 488 mm.

The salient features of the new concept:

- Increased length of the seating bay to increase seating comfort of passengers and facilitate easy movement between the berths

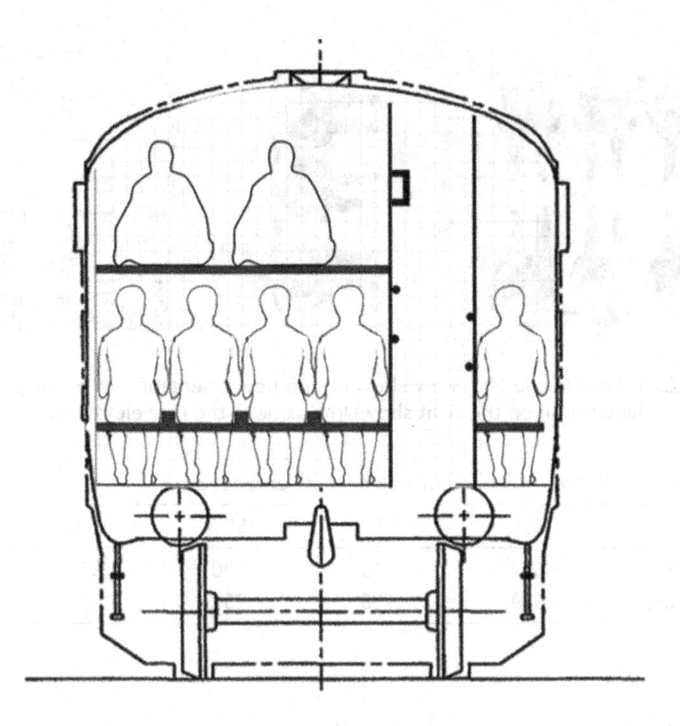

Figure 5.22 Elevation of the compartment showing seating arrangement: concept 1.

- Division of seats into four parts separated by a bump of 80 mm height for providing personal seating space for passengers
- Upper birth can be used for comfortable seating of two persons

5.3.6.2 Concept 2

Default dimension of the space is 3040 mm (cannot be changed).

The default dimension of the space is 3040 mm which must be kept intact. There is only possibility of changing the orientation of the seating layout, which has been suggested. For calculating the space between passengers seated in a sociopetal seating arrangement, the anthropometric dimension of relaxed seating was taken.

5.3.6.3 Relaxed seating

To ensure relaxed seating in Figure 5.23, the individual personal space needs to be calculated. If we see the space from the plan, then the part of the body which protrudes out maximum are the legs (Table 5.11). In elevation this is conformed further.

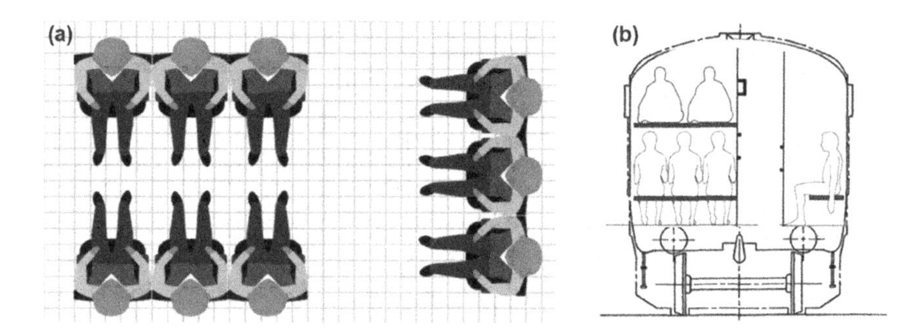

Figure 5.23 Layout in the plan view showing the new orientation of passenger seating. Illustration on the right shows the same in the rear elevation.

Table 5.11 Relaxed seating/buttock to knee extended

	5th	*25th*	*50th*	*75th*	*95th*
Male	652	765	820	842	999
Female	610	708	752	802	899

This is an issue of the clearance. The challenge is we cannot increase the overall space of the compartment because that is the technical constraint. The available space for sociopetal sitting arrangement for two sets of seats is 1768 mm. We cannot go by the highest value here as it would exceed the space. So, the optimized value with which we must play is 75th percentile male value, i.e., 842 m. So, drawing an arch with 842 mm on either side of the seat provides a space of 1684 which is within 1768. Thus 75th percentile users can sit with ease, but 95th percentile passengers must compromise on their comfort. In the same space, another row of seating was arranged (Figure 5.23) facing the aisle to ensure comfort.

Salient features of concept 2 layout

There was an increase in the aisle space compared to before (Figure 5.24). This led to comfortable seating for nine passengers in one bay with different percentile anthropometric dimensions. The upper berth suggested here could be used for comfortable seating for two passengers.

5.3.6.4 Final concept

This concept is an extension of concept 2 (Figure 5.25). The reason for selecting this concept is that it increases the space between the berths and the aisles. It facilitates the seating for passengers with different somatotypes

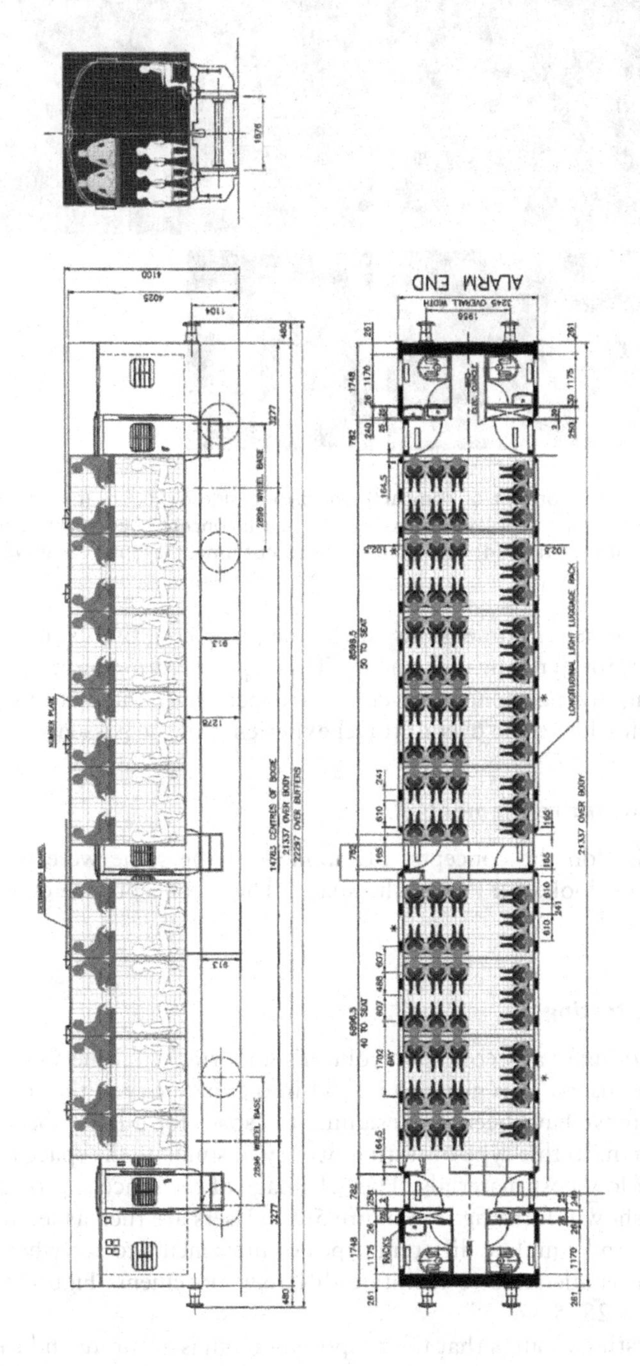

Figure 5.24 Seating layout for concept 2.

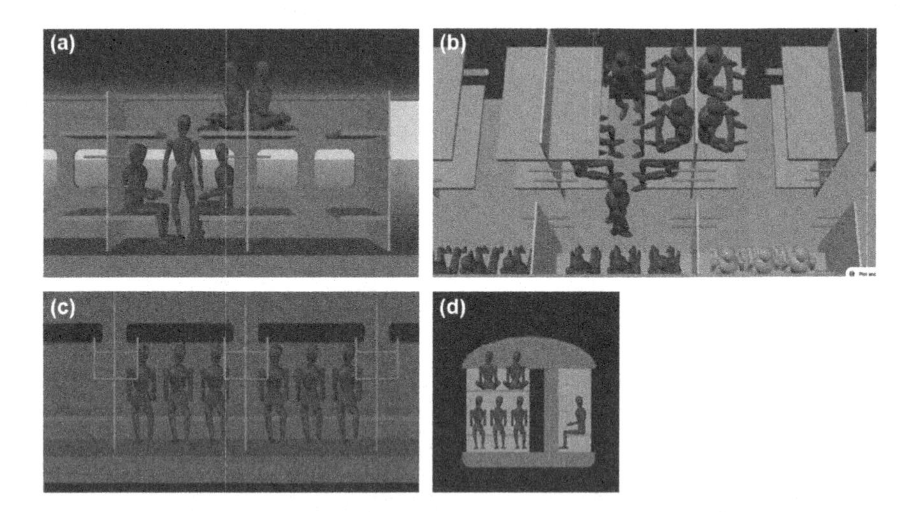

Figure 5.25 (a) Final concept of compartment layout: side elevation. (b) Final concept of compartment layout: plan view. (c) Final concept of compartment layout: front elevation. (d) Final concept of compartment layout: rear elevation.

and anthropometric dimensions with adequate comfort as it provides enough space for dynamic movements. This type of seating promotes sociofugal seating arrangement, thus enhancing passenger interactions during travel, which adds to the overall travel experience of the passengers.

5.3.6.5 Concept refinement

After freezing on the concepts, 3D models of the same were generated to get a better look and feel of the space. These concepts are depicted in Figure 5.25a–d.

5.3.6.6 Rig testing

After converging to any concept layout, it's important that the same is validated on real users. This gives an insight into dynamicity in the space, and how accurate we have been, in designing the space for a large spectrum of the population. In this type of testing, we try to simulate the space through the usage of low-cost materials. The 1:1 image of the space is projected on a screen as shown in the figure (Figure 5.26). Users are then asked to move in the space and simulate different types of movements to see whether the space design is adequate to facilitate different movements Figure 5.27a–c and Figures 5.28–5.37.

The rig testing ensures that the proposed design is accurate and caters to the dynamic movement of users in the space. The next step is making the

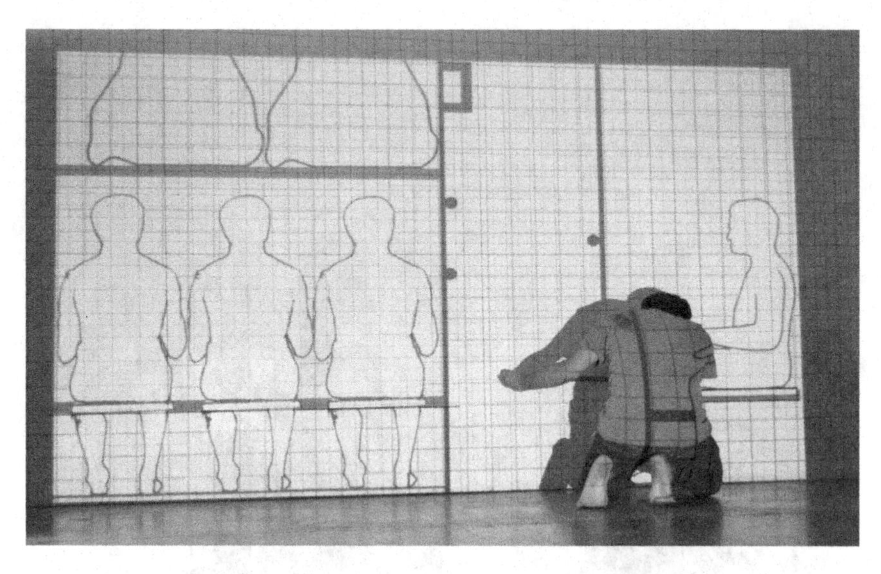

Figure 5.26 The space being projected (1:1) against a grid board for ease of calculation. The seating arrangements are shown so that users can simulate the same movements. All the elements in the space are to scale.

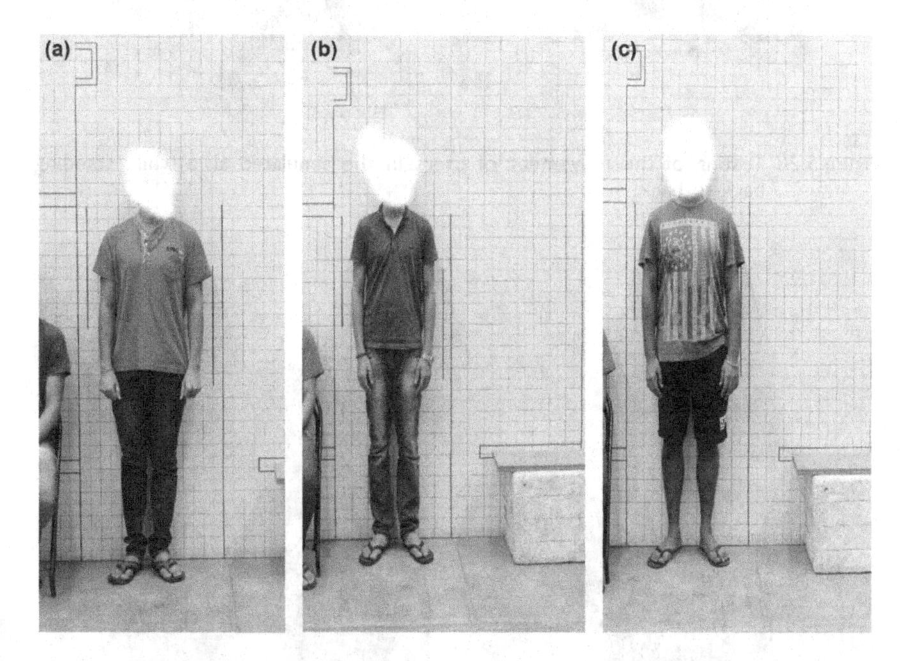

Figure 5.27 (a) Aisle widths being tested for users with varied anthropometric dimensions: participant 1. (b) Aisle widths being tested for users with varied anthropometric dimensions: participant 2. (c) Aisle widths being tested for users with varied anthropometric dimensions: participant 3.

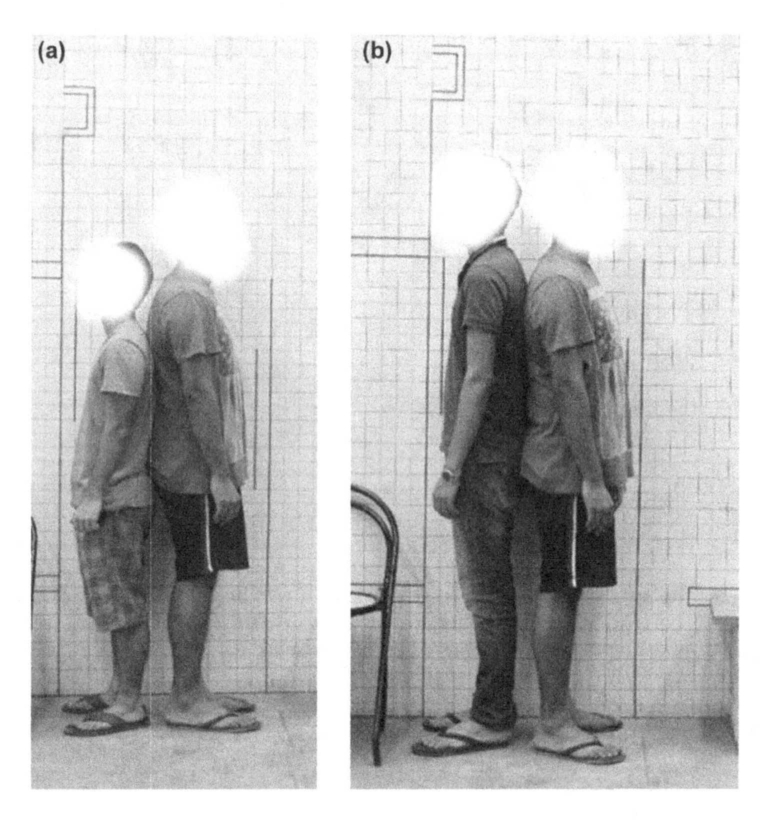

Figure 5.28 Testing of the movement of users in the simulated aisle while standing back-to-back.

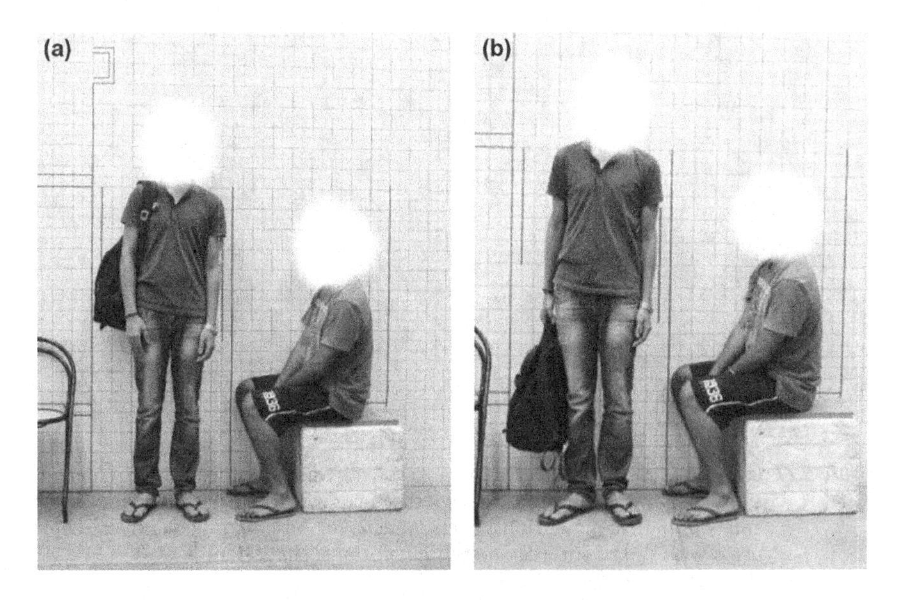

Figure 5.29 Testing of user's movement with baggage in different ways.

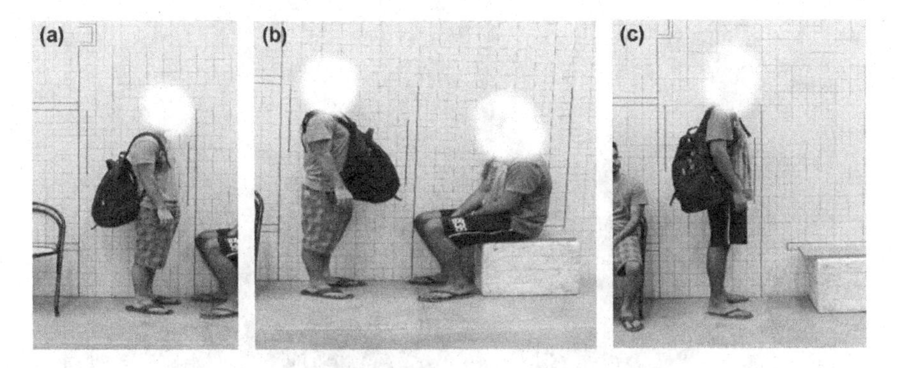

Figure 5.30 Testing of users of different dimensions moving with baggage in different ways.

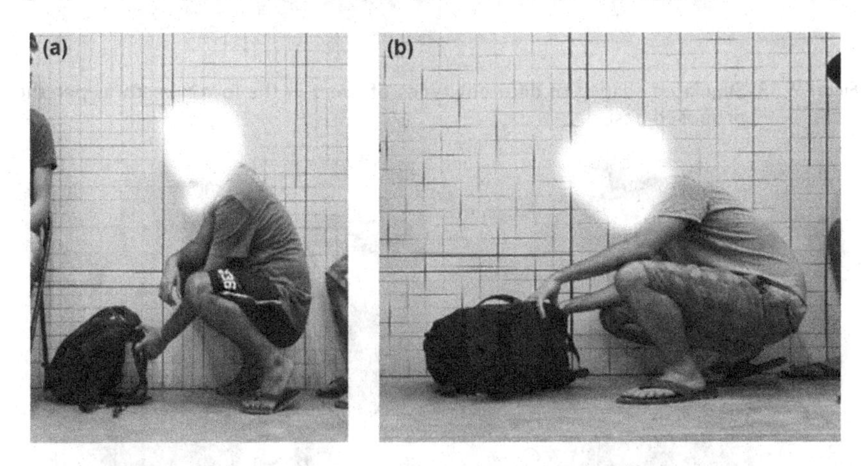

Figure 5.31 Testing for users accessing baggage below the lower berth.

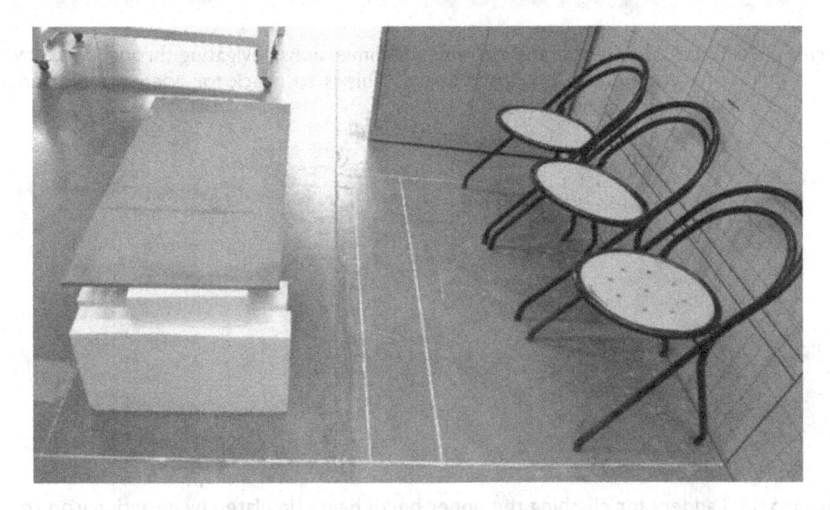

Figure 5.32 Rig for mimicking the lower berth of the compartment. Metallic chairs, polystyrene sheets, and medium density fiber (MDF) board are used as props.

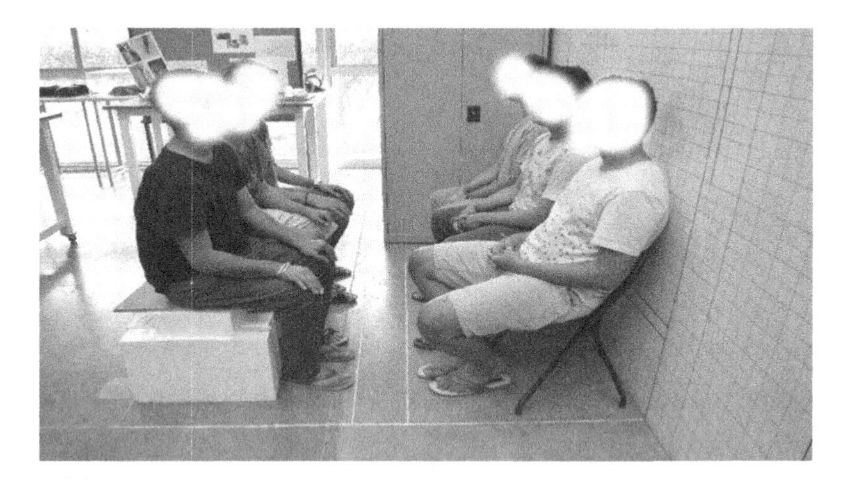

Figure 5.33 Simulated seating of different types of users in the lower berth as per the proposed design.

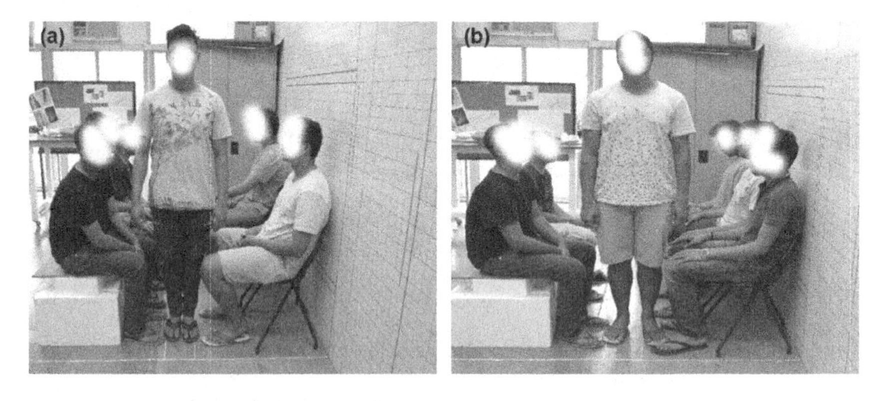

Figure 5.34 Users of different anthropometric dimensions navigating through two rows of seating in the proposed space. This is to check for adequate clearance when the users are moving.

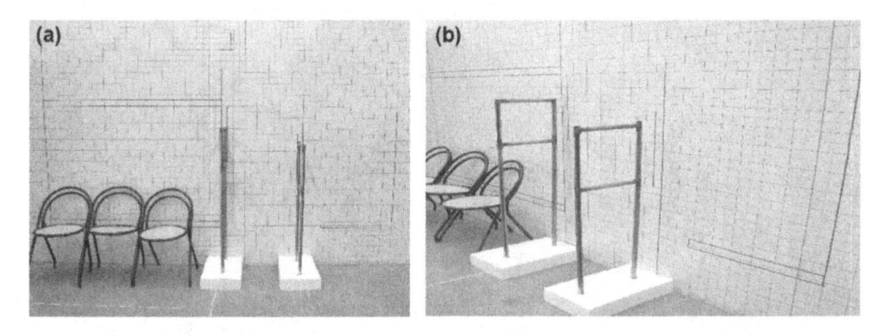

Figure 5.35 Ladders for climbing the upper berth being simulated by inserting iron rods into thick polystyrene sheets.

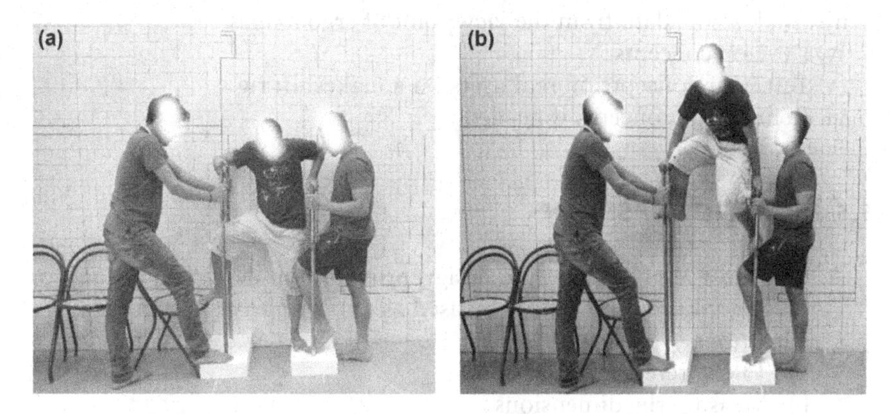

Figure 5.36 Users climbing the upper berths using the proposed dimension of the ladder. To ensure safety while climbing, two assistants are present.

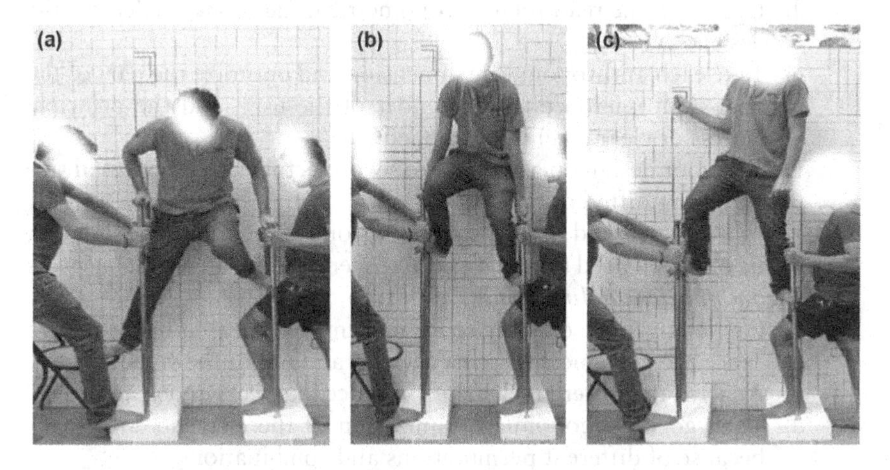

Figure 5.37 Users climbing the upper berth using the ladder and grabbing the handle fixed on the other wall of the compartment.

actual prototype of the space and testing it out, which is expensive and follows the same process of testing.

5.4 KEY POINTS

While going for ergonomic design intervention in any moving space, the steps to be followed are as follows:

i. Select the methodology for data collection
ii. Collect data specific to the space and the type of ergonomic intervention you are looking for

iii. Analyze the data from the viewpoint of ergonomics
iv. Develop concepts
v. Test your concepts on real users on a makeshift rig
vi. Refine your concepts if needed

5.5 PRACTICE SESSION

A. You must apply ergonomic design principles in designing a boat for four people and this would be used as a tourist boat on small rivers
Directions
   a) Take a reference boat
   b) Measure the dimensions
   c) Find out the different touchpoints in the boat with reference to the users
   d) List down the relevant anthropometric dimensions needed for the design
   e) Take each anthropometric dimension and optimize the part of the boat with reference to the gender of the user, and the principle (access or clearance) to be applied
   f) Test your design on a grid board by projecting it to its original size and using different types of users

B. You have been asked to give your views on the ergonomic design of a public bus. You need to focus on passenger comfort and safety aspects
Ergonomic design directions
   a) First analyze the existing space in terms of overall dimension
   b) Trace the movement of passengers in and out of the vehicle
   c) Analyze the different elements in space like seats and grab rails
   d) Then go for ergonomic optimization of the elements and space because of different permutations and combinations

Chapter 6

Ergonomics in communication

OVERVIEW

This chapter gives an insight into the application of ergonomic principles in communication design. Three examples have been selected. The first one is a fire extinguisher, second a pesticide packaging used by farmers, and the third is a map for big college campus to be used by different types of users. The different ergonomic issues, their identification, and ergonomic design directions have been depicted through illustrations and examples. The students should follow the steps to apply the ergonomic principles in their projects.

6.1 INTRODUCTION

Ergonomic principles play a very important role in the domain of communication design. This includes a signage system, graphic design, etc. The ergonomic principles are applicable in determining the placement, size of the boards, color and contrast, font size, etc. Application of ergonomic principles could be done along two lines: physical ergonomics in terms of dimensions of information system and visual ergonomics which exclusively deals with ergonomic issues in typography, color and contrast, font size estimation, text alignment, searching time, etc.

6.2 SOME DIRECTIONS FOR ERGONOMICS IN COMMUNICATION DESIGN

Ergonomics in communication design mainly follows two paths. The first one involves physical ergonomic aspects dealing with a visual cone, such as the distance between two sets of information in a space. The exclusive visual ergonomic part deals with different visual ergonomic issues like text alignment, foreground and background color, font family and size, and use of all caps and upper-/lowercase.

DOI: 10.1201/9781003302933-6

6.3 FIRE EXTINGUISHER

This exercise shows the application of ergonomic principles (Figure 6.1) in designing labels for equipment which are used in emergent situations. Here the focus should be on the context of use as well as the varied group of users.

The following steps could be followed to arrive at the proper ergonomic solution to the problem associated in communication with the equipment.

a) Location of the device: It's important to study where exactly the device is located to get an insight into the context of its use

Observation revealed that one extinguisher was attached to the wall and two were lying on the floor. Further the letters capital A, B, and C were written which however made no sense to the users.

Figure 6.1 Location of the equipment on the wall of the building.

Figure 6.2 Visibility of the information and movement of the eyes.

b) Visibility of the information on the equipment

User study with the information label (Figure 6.2) revealed that the users are confused about the sequence in which the label is to be read. The focus first goes on the capital A, B, and C and then on the branding. The curvature of the equipment surface makes it difficult to scan the information in one go as much information is hidden at the end of the labels because of the curvature.

c) No clue about the type of extinguisher: users were confused as to the type of extinguisher and what type of fire could be handled by that type of equipment
d) No guidance on how to operate the equipment in the event of a fire
e) Users were not aware of the precautions to be taken while using this. The information for the same was too small
f) The label had no consistency in terms of its placement on the equipment
g) The capital letters A, B, and C which convey important information were also inconsistent as seen in Figure 6.2

Based on the above observations and user study feedback, a new ergonomically designed label for the equipment was made with the following features and the steps followed were as follows. Figure 6.3 was the information breakup for different context on the fire extinguisher.

Figure 6.4 was the first "how-to-use" sticker with the following ergonomic features:

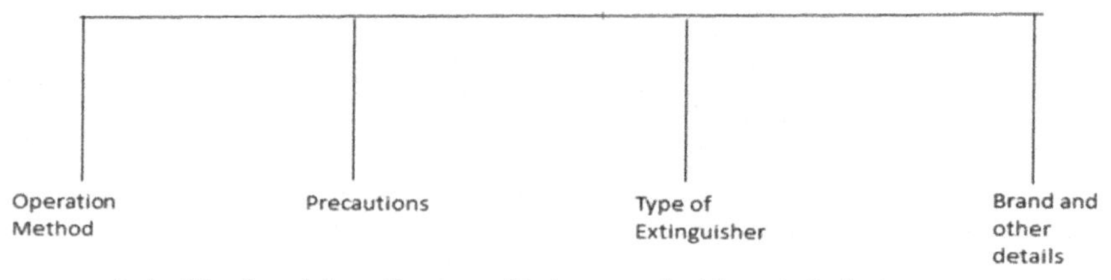

Figure 6.3 Hierarchy of information based on the user study for operating the extinguisher.

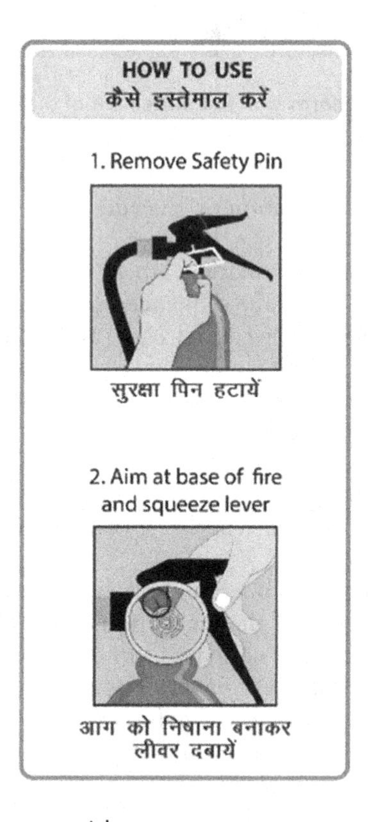

Figure 6.4 The first how-to-use sticker.

a) Usage of product in tandem with users' mental model, and the equipment parts were illustrated to mimic the original equipment
b) The usage was reconfirmed through the vertical arrangement of the label (top to bottom as one reads) which is a neutral stereotype while reading
c) The reading sequence is further reconfirmed through the prudent usage of numbers to show its sequence in which it is to be read
d) The context of the text was minimized to reduce cognitive load while reading the label

The precautions sticker (Figure 6.5) had ergonomic features as they were presented in the order of priority from top to bottom. Minimum text was used so that reaction time is faster. Visuals were used after testing them for comprehension with the users, and they were supplemented with minimal text for reconfirmation.

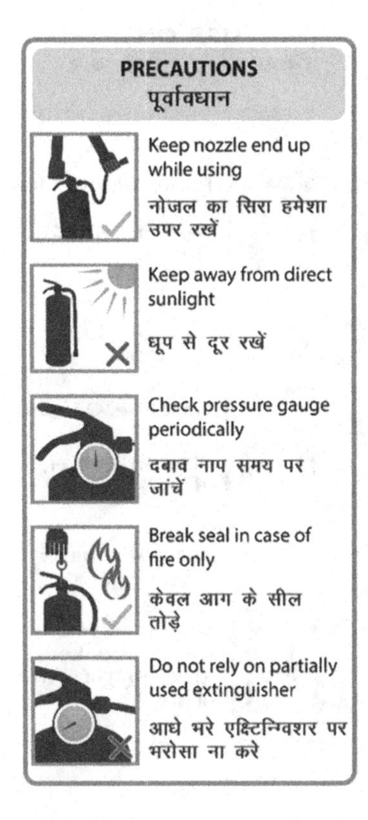

Figure 6.5 Precautions sticker.

The next sticker was the use-on sticker (Figure 6.6) which indicates the type of fire on which the equipment could be used. The size of the letters A, B, and C was reduced, and more emphasis was on the visuals indicating the type of substance on which the extinguisher could be used. The visuals were color coded in green, red, and yellow to further reemphasize their usage.

The sticker for branding (Figure 6.7) is clear and less distractive. It would lie close to the wall and is not that important while the product is being used. The branding is clear and conveys a clear message to the target users.

The stickers were also tested in context, as shown in Figure 6.8. All the information is presented to the users in a very easy-to-understand manner in tandem with visual ergonomic principles.

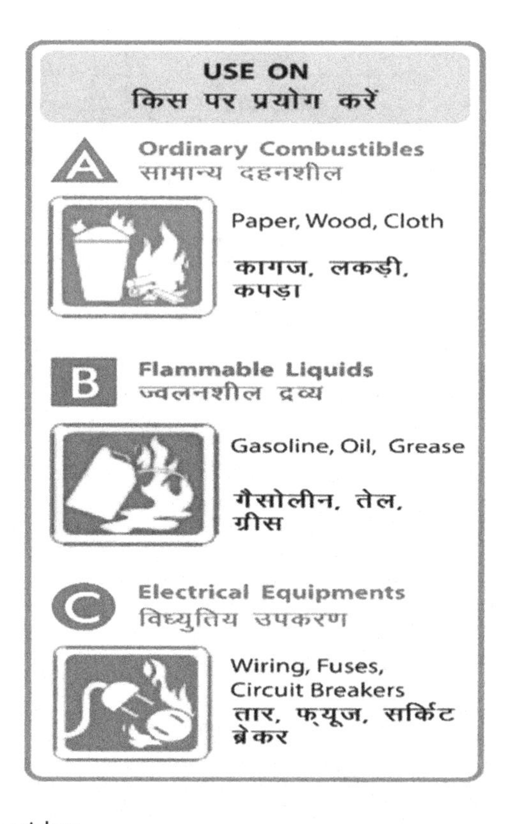

Figure 6.6 Use-on sticker.

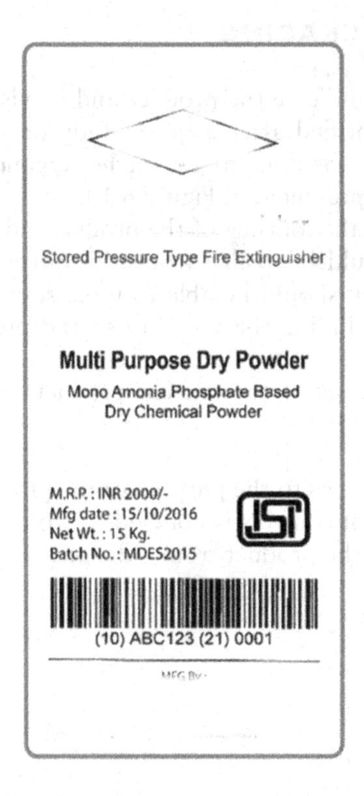

Figure 6.7 Sticker for the brand.

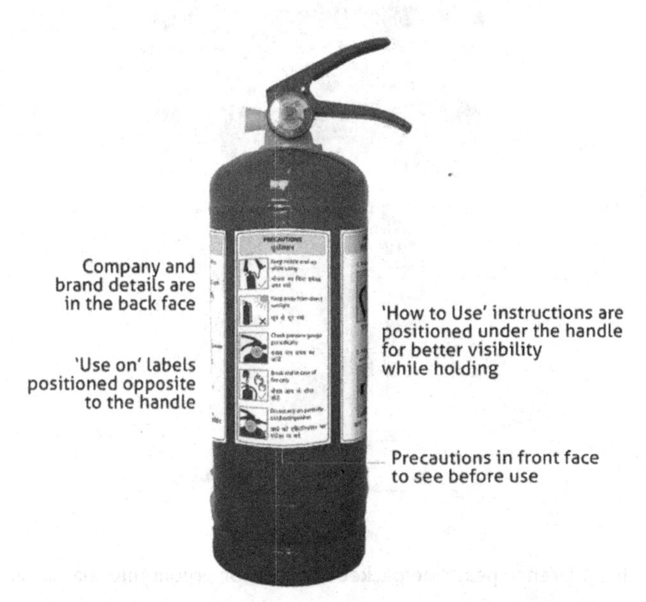

Figure 6.8 Final information stickers on the equipment in context.

6.4 PESTICIDE PACKAGING

Figures 6.9 and 6.10 indicate the product and its visual ergonomic issues, which have been identified after a gross study on different target users. A systems-level study was done to get further ergonomic insights into the packaging which is represented in Figure 6.10.

A detailed study of the journey of the product indicates that as it enters the shops then it should be visible from a distance. When it reaches the customers' house, they should be able to understand all that the packaging has to convey, including the way to use and precautionary measures (Figure 6.11).

An ergonomic analysis of the current product indicates the following issues:

a) Users have no clue as to the purpose of the product
b) The type of pest it controls is not evident from the product
c) The visibility of the product from a distance for example in a shop is very poor

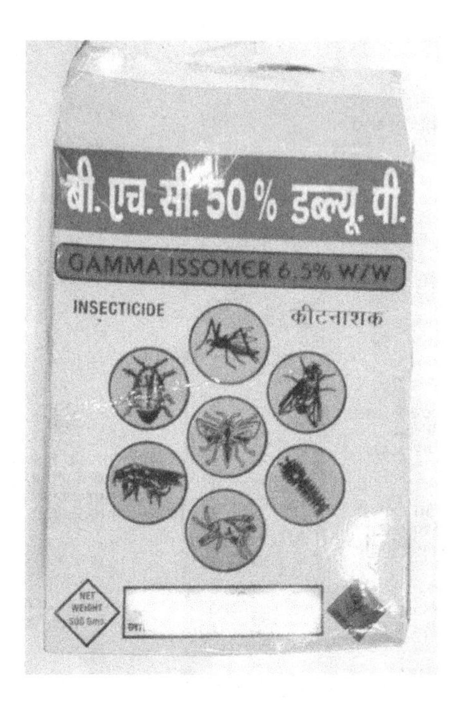

Figure 6.9 The reference pesticide packet selected for ergonomic analysis and design.

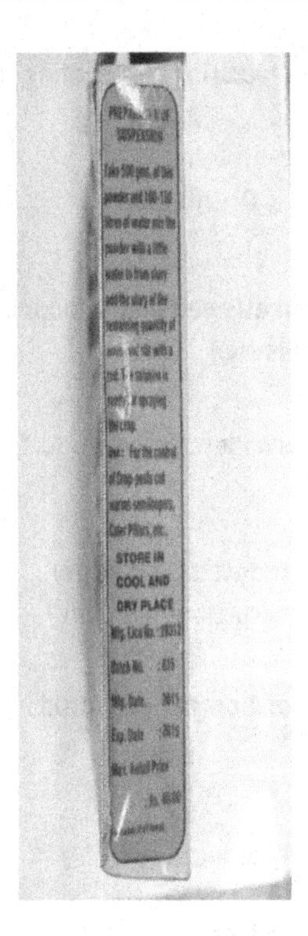

Figure 6.10 Side of the packaging.

d) Under low levels of illumination in small shops in rural areas. the product does not effectively convey its contents and usage
e) The graphics are not designed in tandem with the mental model of the users, the majority of whom are illiterates or semi-literate

Thus, with an eye to the above ergonomic issues, the following concepts were developed (Figure 6.12a and b). The proposed concept of the pesticide packaging in English is represented from front in Figure 61.2a. The same from the side is represented in Figure 6.12b.

The key features of the new design were as follows:

a) The title of the pesticide is in all caps to draw attention
b) The exact usage of the product is indicated graphically in accordance with users' mental model
c) The logo indicates it's poison and needs care to handle is reconfirmed through the usage of a symbol along with text (symbol of skull and bone)
d) The textual materials are in English and the local language as well
e) On the sides, visuals have been added indicating their intended way of disposal (Figure 6.13)

Figure 6.11 The journey of the product.

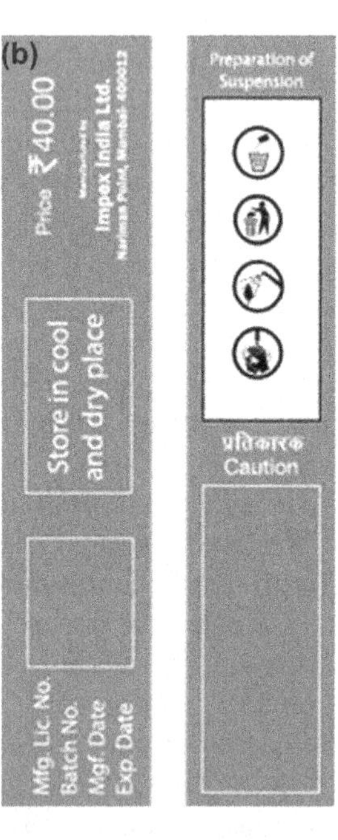

Figure 6.12 Ergonomic design of the packaging.

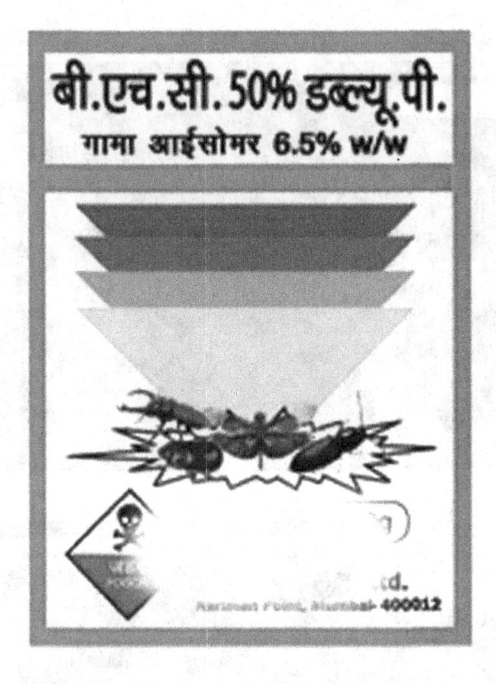

Figure 6.13 Local language being used on one of the sides.

6.5 WAYFINDING/MAP FOR A COLLEGE CAMPUS

A college campus was chosen for designing its map and helping people in navigating to different parts of the campus through the prudent usage of the information system available. Few scenarios were developed to come up with ergonomic solutions for the same.

a) In the first stage, the satellite view of the campus was studied in detail to get to know the position of different buildings and their distance from one another (Figure 6.14a) Figure 6.14b represents the satellite view of the campus showing the academic area in detail.
b) New two personas were created to get an insight into the needs of the users moving in the space

Persona 1

- Parent/Guardian
- Age: 52
- Goals:
  - Submit the application for his child in the academic office

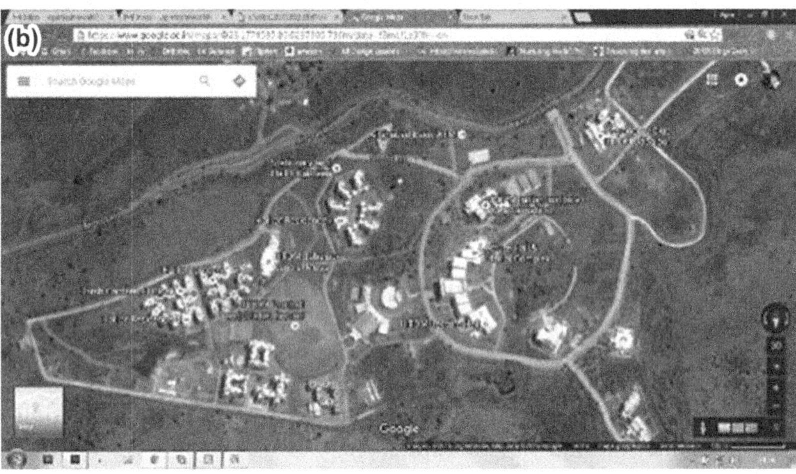

Figure 6.14 Satellite image of the campus on Google maps.

- Talk with administrative staffs
- See the student's hostel
- Description:
  - User is a parent who wants to submit the application form of his child in the college office. He also wants to talk with some staffs to know more about the college. He wants to see the hostel facilities also if possible

Persona 2

- Student from another college
- Age: 20
- Goals:

  - Find venue of sports event
  - Buy food/refreshment
  - Use washroom for cleaning up
- Description:
  - User is a student from another college coming here for an inter-college sports meet. He wants to find the venue of a sports event. Also, he wants to buy some food or refreshments before/after the event. After the event, the user wants to find the washroom to clean up and change

c) In the next stage, the different buildings on the campus were segregated based on different categories as shown in Figure 6.15

d) In the next stage, the color coding was selected to identify each category, as shown in Figure 6.16 depicts different coding used for different areas of the space.

e) After this, the categories of buildings were categorized and prioritized (Figures 6.17–6.19).

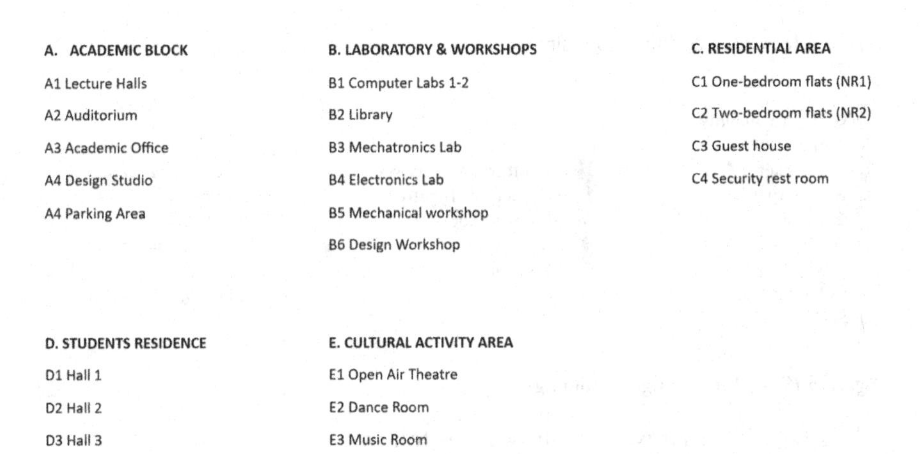

Figure 6.15 Different buildings segregated based on different categories.

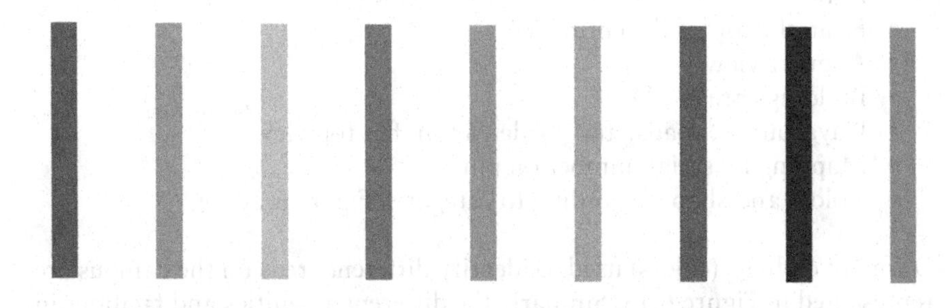

Figure 6.16 Different color coding selected for different categories of buildings.

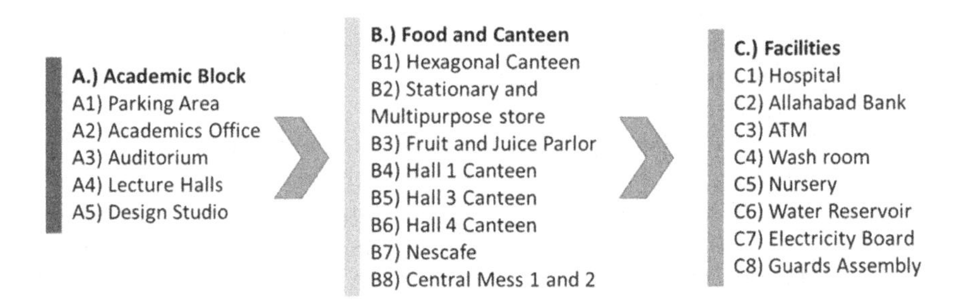

Figure 6.17 Categories of building and their color coding.

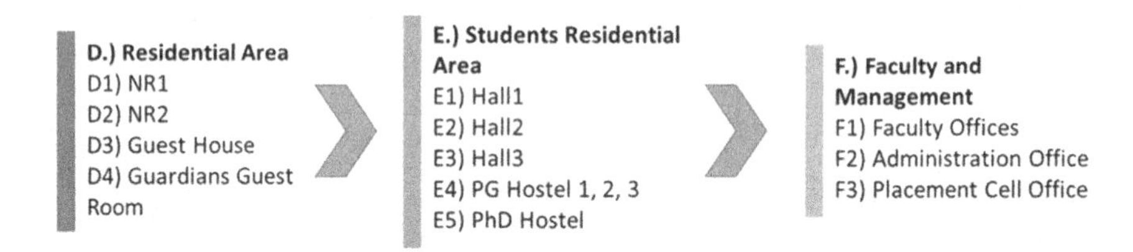

Figure 6.18 Color coding of buildings.

Figure 6.19 Color coding of buildings.

The key points that were applied are as follows:

- Not following magnetic North (new directions: up, down, left, and right)
- From the top of the entrance
- Angular view
- Buildings are in 3D
- Playgrounds, roads, and garden are in flat top view
- Mapping by serial number on pin
- Colors and shape according to category (Figure 6.20a–i)

Different codings (colors) used to identify different areas on the campus are represented in Figure 6.16. Similarly the different amenities and facilities in

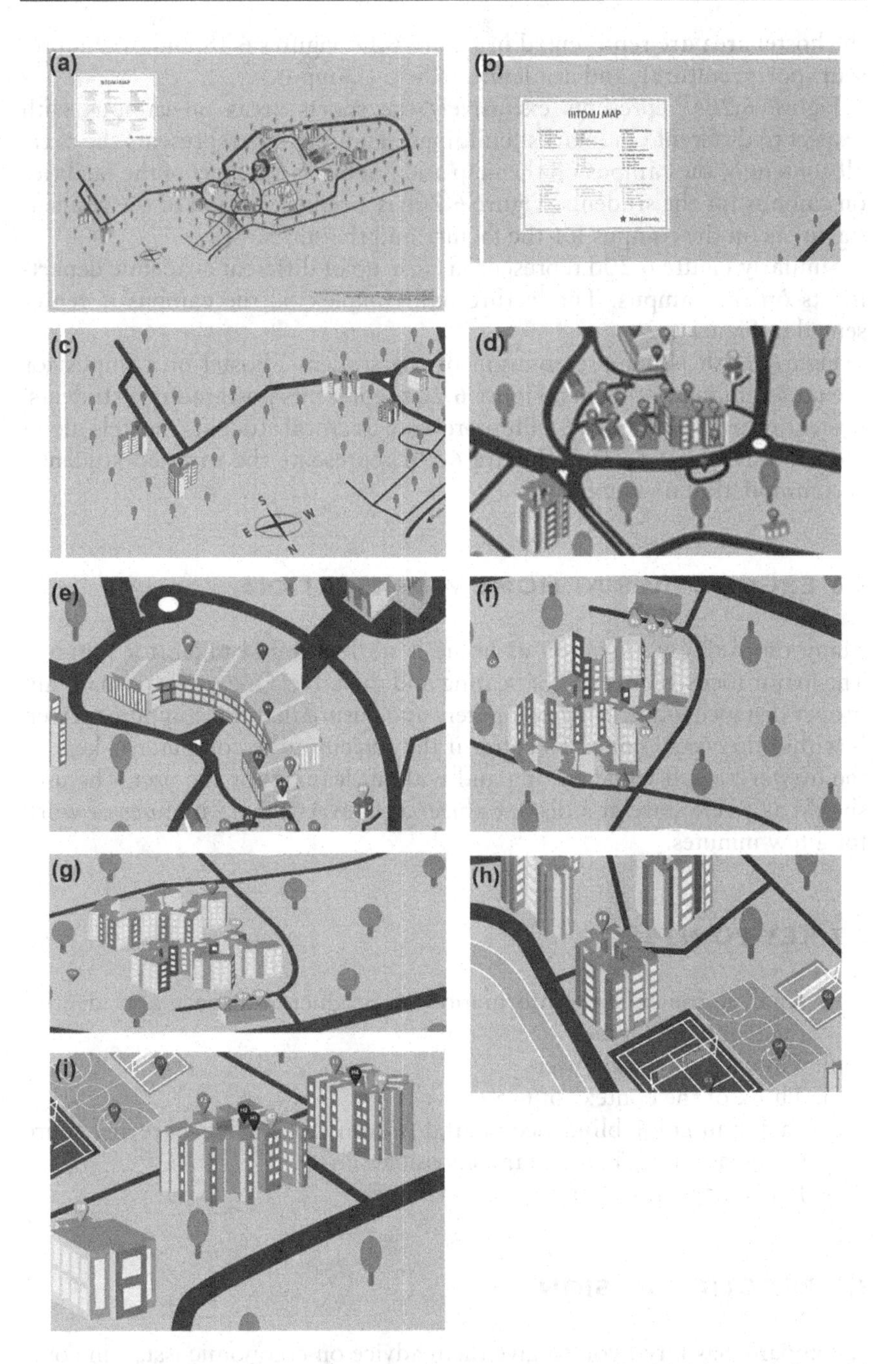

Figure 6.20 (a) Map of campus. Different buildings segregated, and color coded on the institute map. (b) Legend for the campus map. (c) Map with some landmarks. (d) Map with buildings. (e) Map with the academic block. (f) Map with the administrative block. (g) Map with the student hostel. (h) Map with the sports facility. (i) Map with different types of hostels.

the hostel area are represented in Figure 6.17. Figures 6.18 and 6.19 represent sports, cultural, and academic areas on campus.

Figure 6.20a represents exclusively the sports areas on campus with respect to different landmarks (buildings). Figure 6.20b represents the overall content of the campus on the map and is a bird's eye view of the facilities on campus for the students. Figure 6.20c is a representation of the residential areas on the campus for the faculty and the staffs.

Similarly Figure 6.20d represents a close up of different academic departments on the campus. The lecture hall complex on the campus is represented in Figure 6.20e.

Figure 6.20f is a representation of the students' hostel on campus for the undergraduate students. Figure 6.20g represents postgraduate students' hostel on campus. Figure 6.20h represents doctoral students' hostels along with the badminton court. Figure 6.20i represents the married students' accommodation on campus.

6.6 ERGONOMICS AT HOME WORKSTATION

Many times the workstation at home is neglected and taken for granted. The major focus is only on the seating and the table. One needs to maintain proper distance while using computers and ensure that the computer screen is within the visual cone. It's better if the screen of the computer is kept at the forward reach while seating and without leaning for the user. The user should as a rule stare at a distant object after every 30–45 minutes of work for a few minutes.

6.7 KEY POINTS

i. Analyze the existing information on product and space and identify the ergonomic issues
ii. Think of the target users
iii. Think of the context of use
iv. Factor in color blind people and refrain from using pure red, pure green, and pure blue as far as possible
v. For signage, think of bidirectional ones

6.8 PRACTICE SESSION

A company has hired you to give them advice on ergonomic issues in communication aspects of their newly launched biscuit packet. They would like to advertise these biscuits on the roadside as well as the dividers. How would you proceed?

Directions for solution:

a) First apply visual ergonomic principles in designing the packet. Ensure they should be visible on the supermarket racks
b) Users look for certain key information on the packet which is to be in the expected quadrant
c) Use proper background and foreground color to ensure optimal visibility under low illumination
d) Once the packet is done, then start the signage
e) Use ergonomic principles in signage and calculate the height by using the common visual cone and principles of trigonometry

Chapter 7

Ergonomics in user interface design

OVERVIEW

This chapter introduces the readers to the application of ergonomic principles in user interface design. The example of a coffee/tea vending machine has been given. Through illustrations, it has been shown how different ergonomic principles have been integrated into the user interface to make it more user-friendly.

7.1 INTRODUCTION

User interface is the bridge or the channel through which information is fed into the machine by human (controls or input channel), and the machine gives output through the output channel or the display (for a detailed discussion, refer to the book *Ergonomics for the Layman*). Ergonomics plays a very important role in the user interface by designing the same in tandem with the needs and wants of users. The ergonomic principles in the interface design are broadly classified as physical ergonomics which deals with the dimensional attributes and the attributes of biomechanics and posture. The other aspect is the cognitive ergonomic attribute which investigates the aspects of the human mind and designing in tandem with how it works.

7.2 SOME DIRECTIONS FOR APPLYING ERGONOMIC PRINCIPLES

For projects on the ergonomic design intervention in user interface design, the following general steps may be followed. You need to remember that all the steps cannot be applied to all types of interfaces at random.

a) In case of a reference product which requires redesign or a new product design, first analyze the interface which exists or which you are to design. You might use a small matrix comprising of different columns

DOI: 10.1201/9781003302933-7

like task analysis, what I need to know, is feedback present or absent, what type of feedback is needed, what can go wrong, and ergonomic directions. You start with the first column for tasks and then fill up the subsequent horizontal columns

b) A collage would emerge, and you would get ergonomic design directions
c) Now apply the principles of physical and cognitive ergonomics for interface design
d) You need to allocate functions at the interface and decide what should go to the machine and what should go to the man
e) Once this is done, apply the principles of physical and cognitive ergonomics
f) Perform user testing of your product
g) Refine your product further before finalization

7.3 EXAMPLE 1: ERGONOMIC DESIGN OF THE INTERFACE OF A DIGITAL BLOOD PRESSURE MONITOR

A digital sphygmomanometer was selected for the redesign of its user interface. Hypertension is a silent killer, and their numbers are on the mount in rural areas. If the semiliterate and illiterate users can measure their blood pressure and become aware, it would save lives. This was a brief given to the research team.

Accordingly, one of the brands of the digital blood pressure monitor was picked up. An ergonomic analysis of the interface was done with a very detailed user study with the target users, and an ergonomically designed user interface was arrived at. The new design was further tested on the target user to refine it further. The reference sphygmomanometer is shown in Figure 7.1.

The following steps were followed for ergonomic analysis and design of the device interface:

a) Ergonomic analysis of the interface
b) Levels of interaction between the device and users
c) User study with doctors, nurses, and target users to get insight into what exactly is needed
d) Detailed task analysis of the device
e) Focusing on all the elements of the user interface, including icons, graphics, and color

Task analysis with the device indicated many gray areas (Figure 7.2) which demanded ergonomic design intervention.

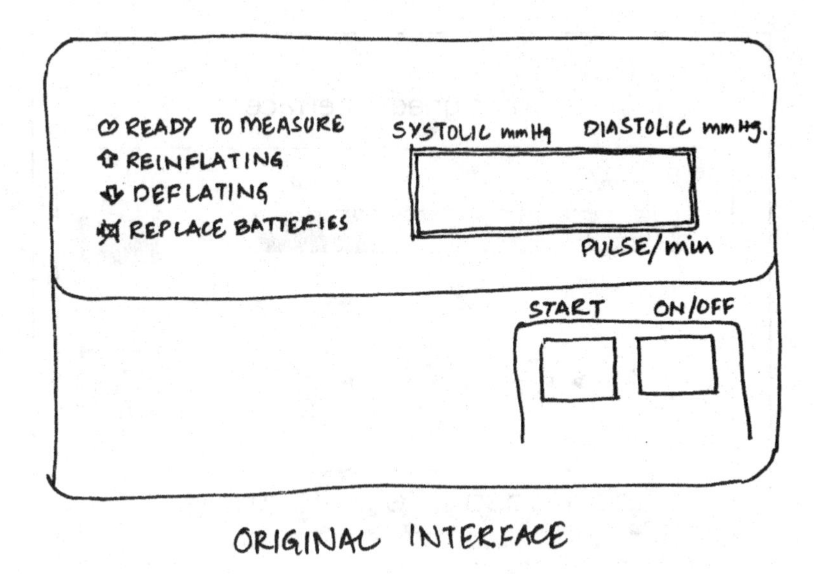

Figure 7.1 Reference digital sphygmomanometer interface used in this study.

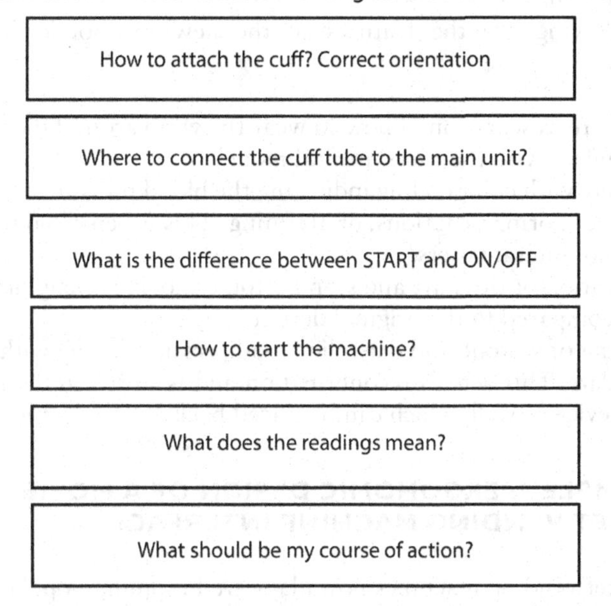

Figure 7.2 Task analysis with the device indicating the different steps and the area of possible ergonomic intervention.

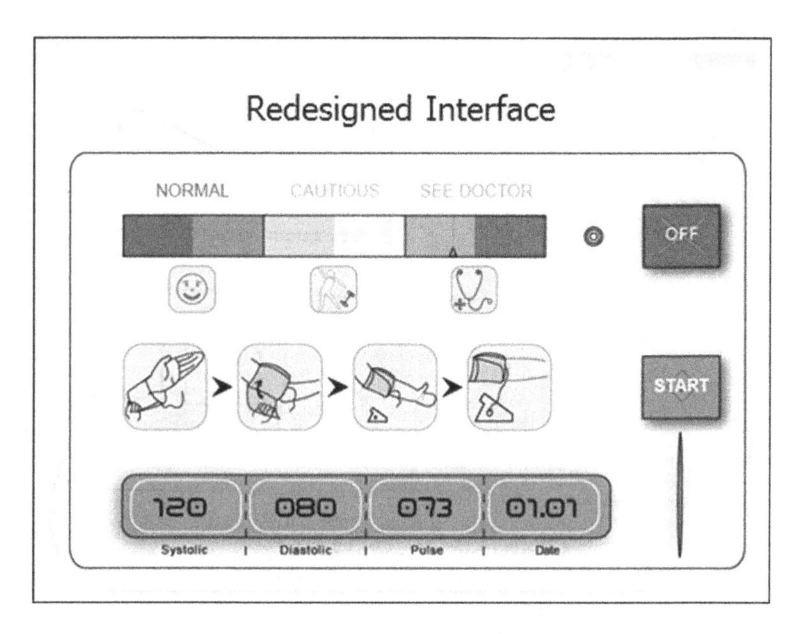

Figure 7.3 Ergonomically designed interface of the device.

Based on the analysis, a new ergonomic interface for the device was designed (Figure 7.3).

The following are the features of the new ergonomically designed interface:

a) Iconic representation of how to wear the wrap around the upper arm, in tandem with the users' mental model
b) Display with color coding indicating the blood pressure and its status, whether normal, cautious, or alarming. This is beneficial to semiliterate and illiterate users
c) Separation of the start and stop button removing ambiguity in operation compared to the original design
d) Labling of systolic and diastolic blood pressures along with pulse rate and date. This acts as reconfirmation and is stored in the memory of the device as well, which can be traced back by the doctor

7.4 EXAMPLE 2: ERGONOMIC DESIGN OF A MOVIE TICKET VENDING MACHINE INTERFACE

Movie ticket vending machines nowadays are becoming popular as people do not like to stand in queues anymore. This project was aimed at making the interface of the movie ticket vending machine much more humane (Figures 7.4–7.6).

Figure 7.4 The home screen of the interface giving an overview to the users.

Figure 7.5 Movie selection interface. Photo by SYLVIA ELIGI from Pixels.

Features of the movie ticket vending machine:

a) Screen one gives a broad overview to the user as to the movies currently going on and the content, cast, and star rating
b) Screen one also gives users information as to the availability of tickets, followed by different options for booking the ticket

Figure 7.6 The kiosk with the interactive movie vending machine.

c) Screen two takes the user to the detail of the movie selected, with a detailed storyline, case, and the option to view the trailer as well. This screen gives various options for purchasing the ticket and even changing one's decision by going back
d) The entire navigation has been designed based on users' feedback and in tandem with the mental model in task performance
e) The screen has been modeled on a kiosk

7.5 EXAMPLE 3: ERGONOMIC DESIGN OF THE INTERFACE OF A COFFEE VENDING MACHINE IN A COLLEGE CAMPUS

The brief given was that a coffee vending machine interface needs to be designed from an ergonomic perspective. Students on the campus of an educational institution demanded self-serving coffee vending machines which they could operate on their own. Based on this brief, an ergonomically designed user interface was designed for a specific group of target users.

Features: based on the user study on the target users, it transpired that the solution should contain the following features:

a) Easy to understand
b) Selection for tea and coffee
c) Indication of price
d) Modality of payment
e) Feedback before deducting the money
f) Location indicator for glass holder

g) Confirmation and reconfirmation at critical stages of operation especially while selecting the number of cups and making the payment
h) Dynamic feedback in the case of drawing attention to the secondary or tertiary quadrant

Figures 7.7–7.24 give a walkthrough of the ergonomically designed user interface.

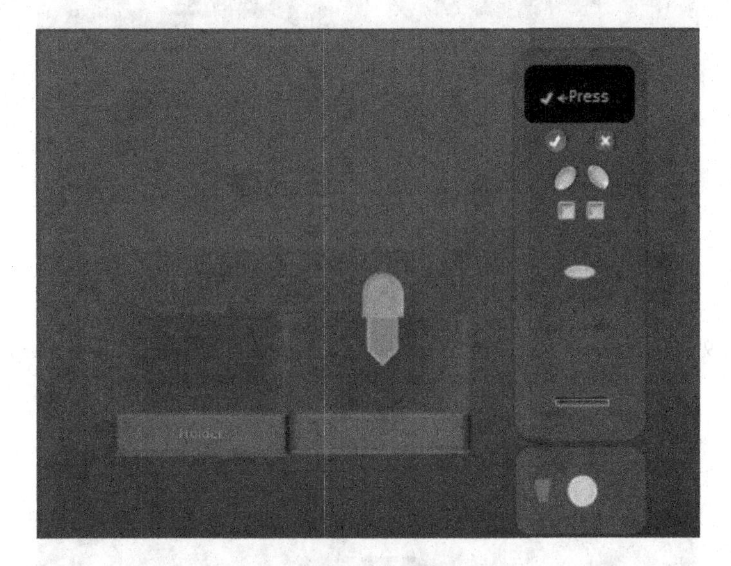

Figure 7.7 Step 1: selection of tick mark to continue machine operation.

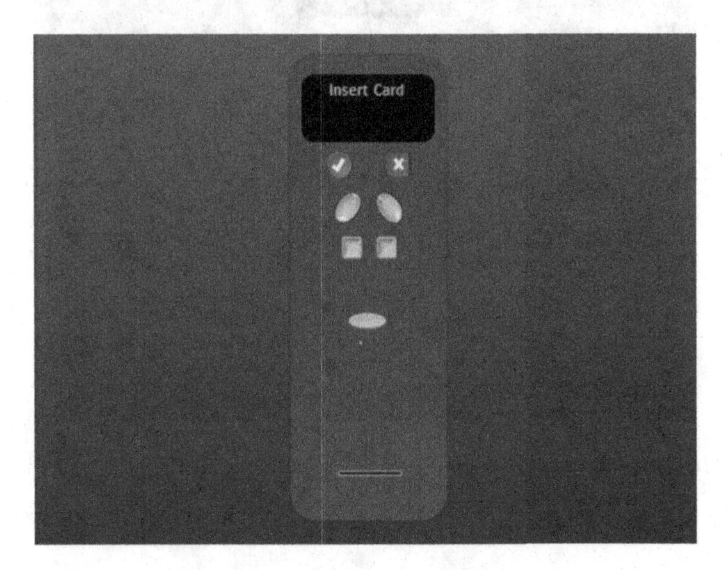

Figure 7.8 Step 2: instructions for inserting the smart card (as proposed based on the user feedback.

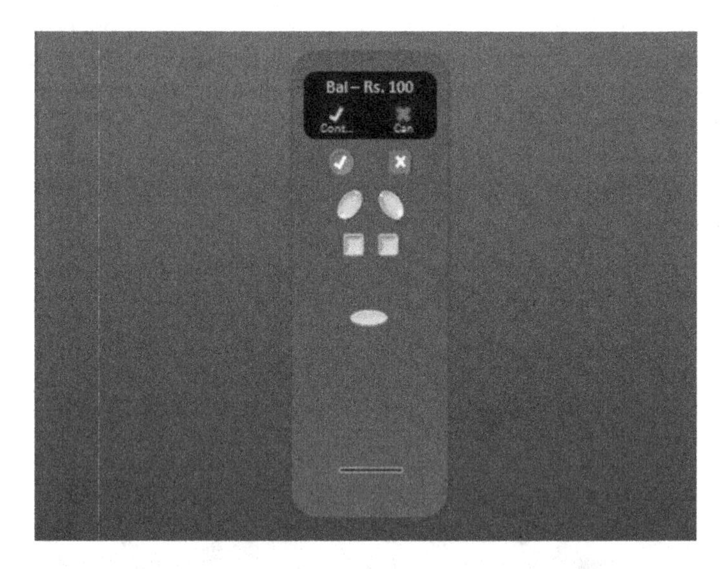

Figure 7.9 Step 3: direct feedback to the users as to their account balance, thus helping them make a decision on how to spend the balance amount.

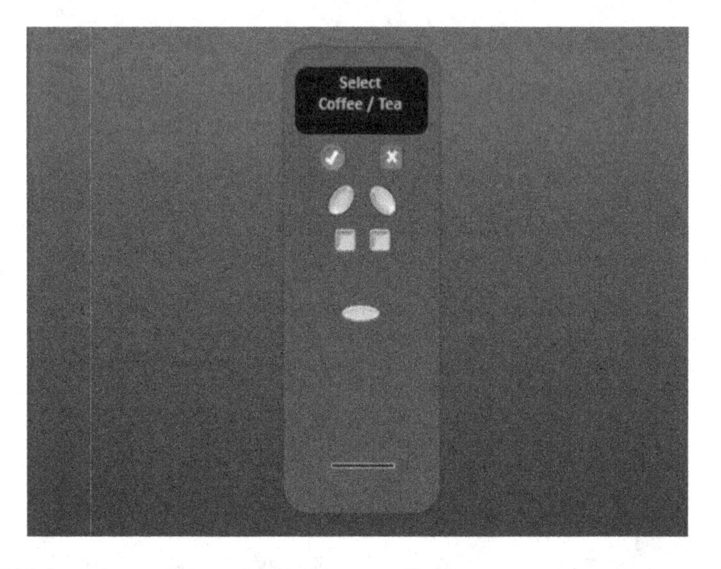

Figure 7.10 Step 4: option for selecting tea or coffee.

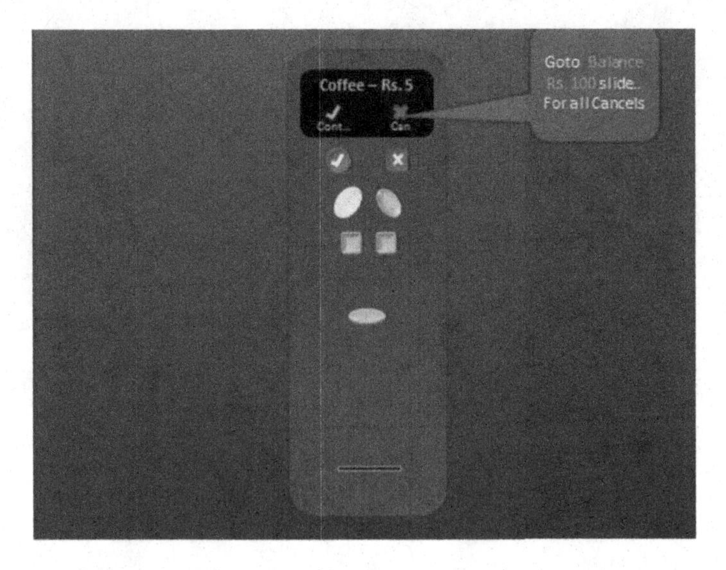

Figure 7.11 Step 5: price of the beverage selected.

Figure 7.12 Step 6: option for selecting the number of cups in case one needs to give a treat to someone.

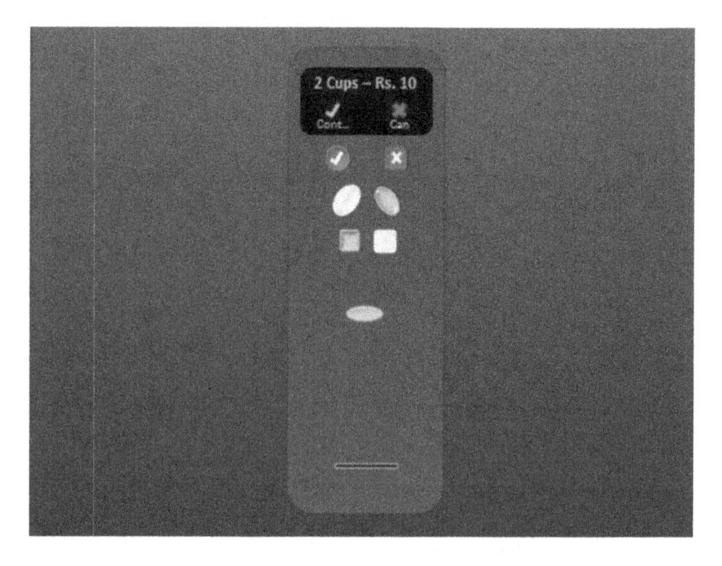

Figure 7.13 Step 7: feedback on the selected number of cups and total cost.

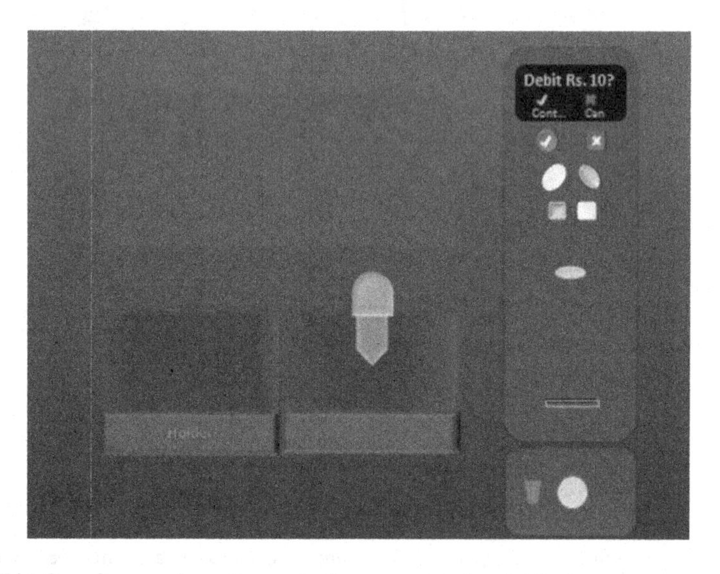

Figure 7.14 Step 8: reconfirmation as to the amount being debited.

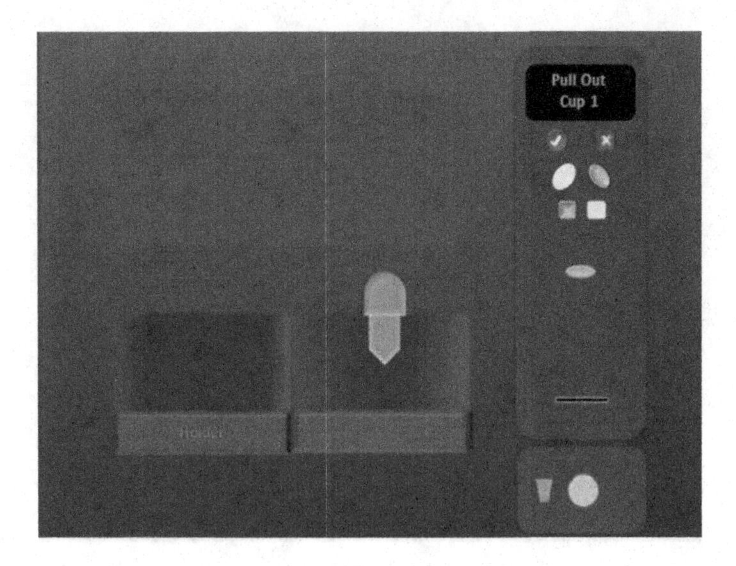

Figure 7.15 Step 9: step-by-step instruction of pulling out one cup.

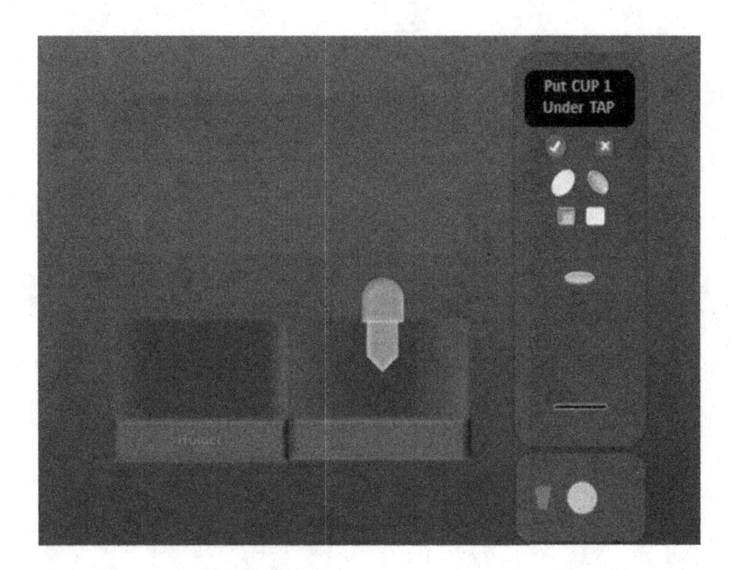

Figure 7.16 Step 10: putting the cup under the tap.

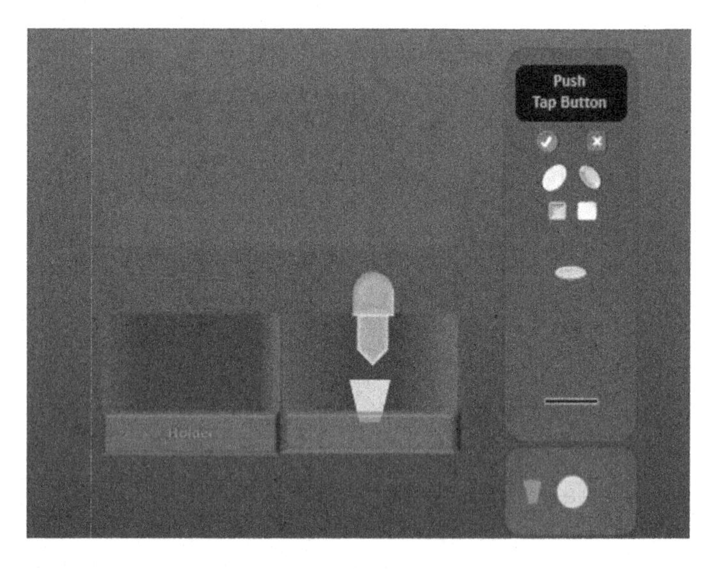

Figure 7.17 Step 11: pushing the tap button for pouring coffee.

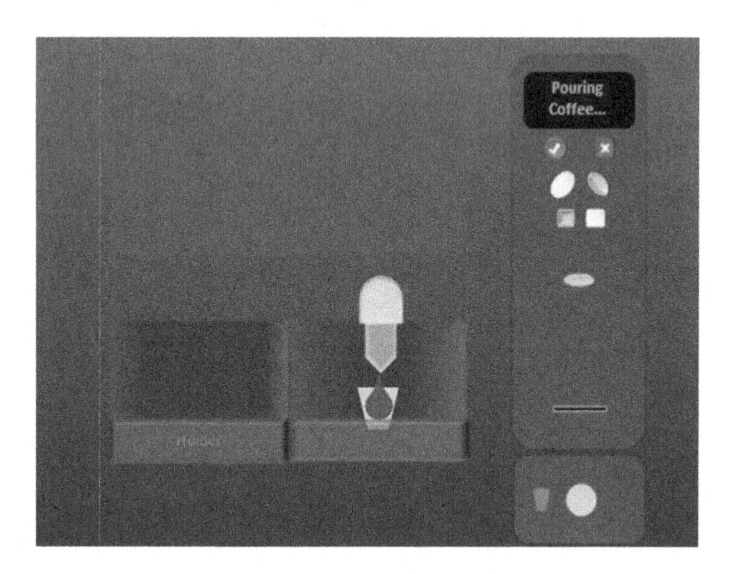

Figure 7.18 Step 12: feedback on the coffee being poured.

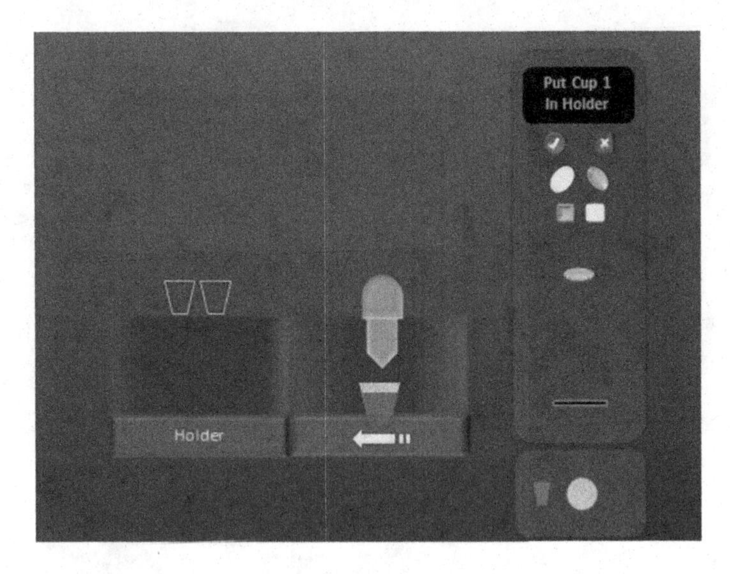

Figure 7.19 Step 13: put the coffee filled cup in the holder so that you can go for the second cup.

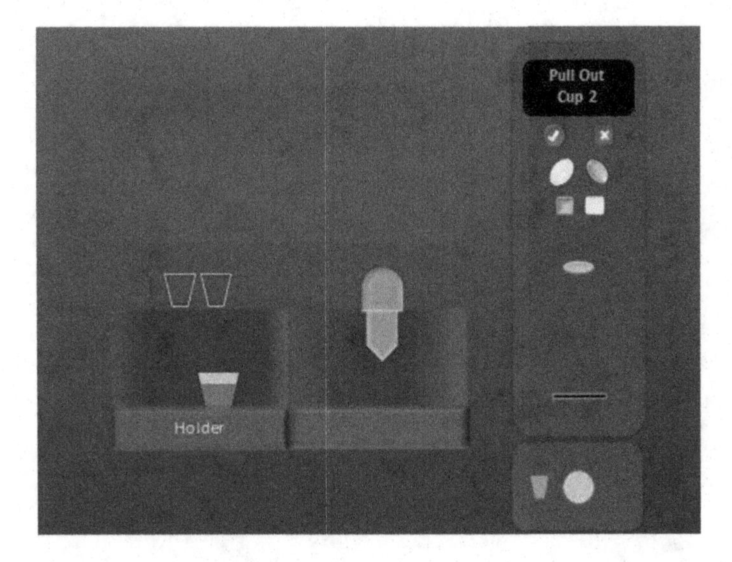

Figure 7.20 Step 14: pull out cup 2.

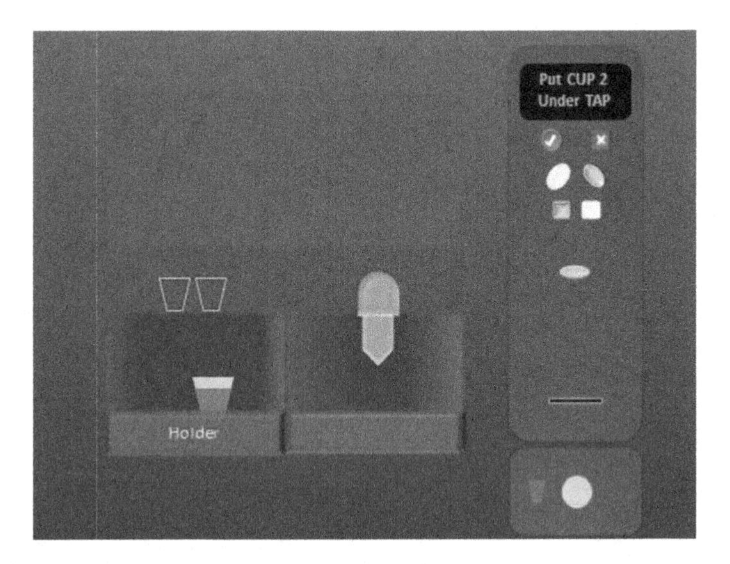

Figure 7.21 Step 15: put cup 2 under the tap as before.

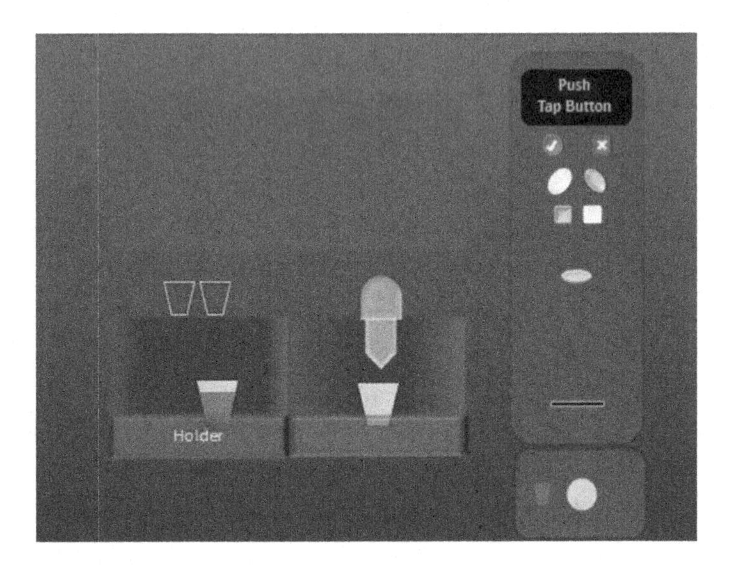

Figure 7.22 Step 16: press the tap button to pour coffee.

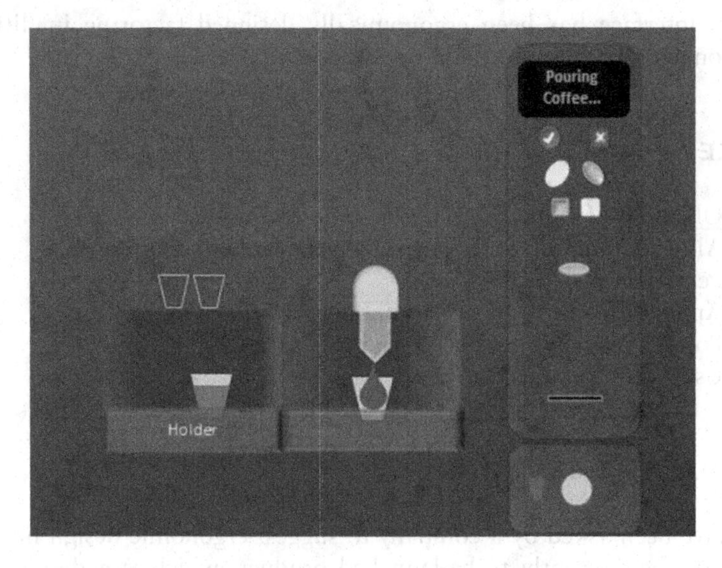

Figure 7.23 Step 17: feedback on coffee being poured.

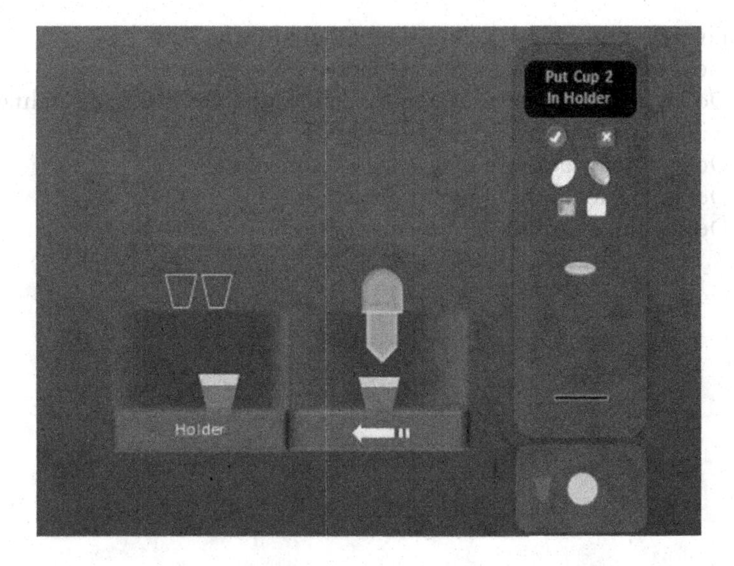

Figure 7.24 Step 18: put cup 2 in the holder so that one need not hold it in hand.

This interface has been ergonomically designed factoring in different ergonomic principles which are narrated in Figures 7.7–7.24.

7.6 KEY POINTS

i) Objective of your design needs to be decided
ii) Allocate function at the interface between man and machine
iii) Perform a task analysis
iv) Apply different cognitive ergonomic principles
v) Population stereotype needs to be factored into
vi) User test your interface

7.7 PRACTICE SESSION

You have been asked by a company to suggest ergonomic design interventions for their recently to-be-launched product, which is a digital menu card. You must suggest ergonomic design solutions for the interface of the digital menu card.

Way forward

a) First do a task analysis in using a digital menu card
b) Next identify the ergonomic issues at every step
c) Do an allocation of function by delineating the tasks the technology will do and those that users would do
d) Define the location of displays and controls
e) Define the location of the information
f) Decide the quantum of information to be presented

Chapter 8

Report and portfolio

8.1 OVERVIEW

After the projects, it's very important that the ergonomic application in design is well documented. Parallel with the documentation of the projects, students should also focus on designing their portfolios in different areas of ergonomics in design.

Any documentation that is done should clearly articulate the problem, methodology, and the output, both tangible and intangible. The outputs should be more highlighted in the portfolio.

For *documentation* of the projects, the following template may be adhered to:

I. Introduction

Here the student should start with the exact area where work has been done with reference to the literature or other similar reports and documents available in the particular area. The writing should start on a broader note and gradually converge into the area which demands investigation. The approach finally will lead to the exact problem statement.

II. Methodology

Here the different tools for collecting data from the end users are discussed. It's not only the tools but also how they were used for data collection is explained here. All the methodologies or methods should be stated with proper references from where they are collected.

III. Results

This section deals with the output of data collected from the end users. The data are properly tabulated for easy understanding.

DOI: 10.1201/9781003302933-8

IV. Discussion

This is where the results need to be explained and their implication in design.

V. Concepts

This section deals with the concepts or solutions generated addressing the problem statement. This section should also discuss in detail the ergonomic design features of the solutions proposed.

VI References

All the books, journals, and documents referred to in this document should be listed in alphabetical order.

For the portfolio (Figure 8.1), one should remember that the emphasis is much more on making it visually appealing. This should contain a brief about the problem statement, the methodology followed, and the

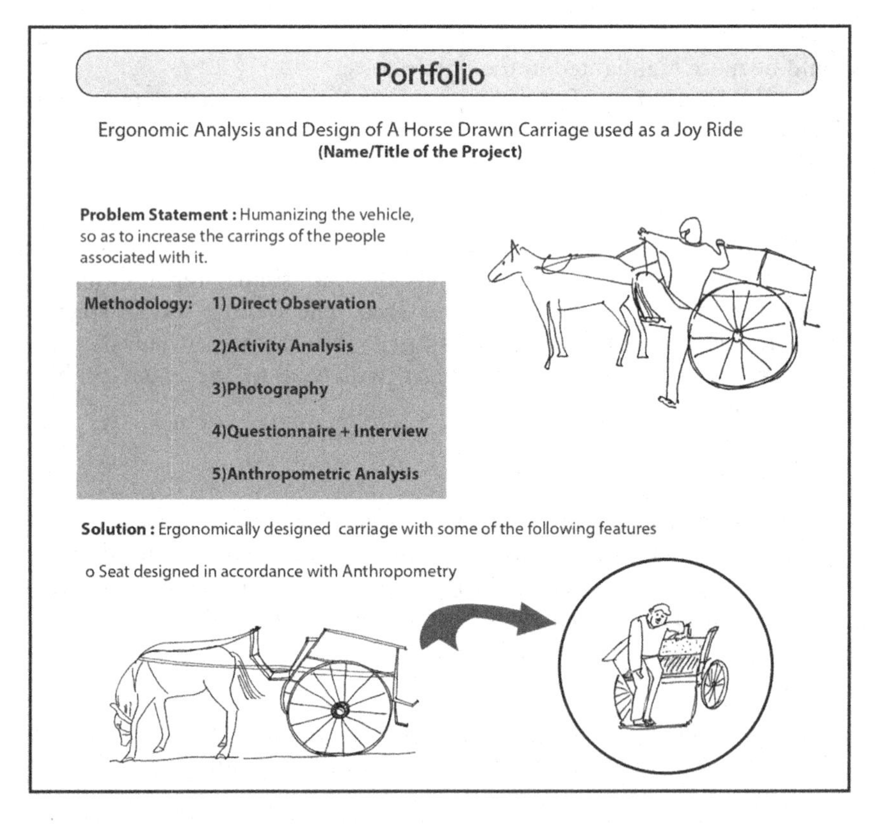

Figure 8.1 Sample portfolio with ergonomics in design assignment.

solutions. Each of these should be represented by visuals from a different perspective so that anyone looking at the portfolio should get an idea of what the problem statement was and what exactly the solutions are, with an emphasis on ergonomic design features.

8.2 KEY POINTS

i) You need to give a bird's eye view of your work done
ii) The problem/opportunity statement needs to be clearly defined
iii) The methodology that you followed needs an explanation
iv) Do not give details of your concepts if it is not protected under Intellectual Property Right
v) Acknowledge all books, documents referred, and all those who helped you during the project

Chapter 9

Application of mannequin in ergonomics

9.1 INTRODUCTION

Mannequins are models of human figures in either 2D or 3D. These models have movable body parts which are hinged. They are normally of different percentile values and are used for testing the ergonomic design while the design is still in a very nascent stage and also when the design is complete. They give an estimate of the accuracy of the design solutions with respect to the entire spectrum of the population.

Another important tool which is used along with mannequins is the anthropometric grid board. These boards are graphs of 5 or 10 sq cm. The mannequins are used on the grid board to check for the different ergonomic design-related issues. This is often followed by the rig test, where very low fidelity prototypes are tested with actual users on the grid board to check for ergonomic design issues. All mannequins used are representative.

9.2 MANNEQUINS OF DIFFERENT PERCENTILE VALUES

You may use mannequins of different percentiles to get a look and feel of the product or space you have designed in terms of its accuracy with reference to the users' touchpoints.

9.3 SCALE DOWN GRID BOARDS WHICH CAN BE USED FOR CHECKING THE DESIGNS EITHER ON A PAPER OR IN THE LABORATORY ON THE GRID BOARD TO SCALE

Below some grid boards are depicted for usage in the classes and in the design studio.

DOI: 10.1201/9781003302933-9

9.4 USAGE OF A MANNEQUIN AND GRID BOARDS FOR TESTING OUT THE ERGONOMIC DESIGN

Both mannequins (scale down or 1:1) act as a good resource for the initial testing of your design, and later on, this could be tested out on real users (Figures 9.1 and 9.2). Figure 9.1 represents usage of the grid board for calculating anthropometric dimensions. Figure 9.2 represents maniquins of different percentile values.

For the sake of the students' practice, some mannequins of 5th percentile male and female and 95th percentile male and female are given (Figures 9.3–9.6). These mannequins are not to scale but are for representational purposes only. Students may make their mannequins out of these templates, which consist of different segments of the body and incorporate actual anthropometric dimensions in them. Even these representative mannequins could be made 1:1 or scaled down after incorporating actual anthropometric dimensions of the target population.

In Figure 9.7 mannequins are given in different orientations like rear and side elevation and plan view. Figure 9.8 is a representation of a grid board in A-3 size, which students can use for designing a space layout with mannequins and if they wish they might do it in a 1:1 scale or scale down it as per their convenience. A copy of the grid can be taken to reproduce it to different scales.

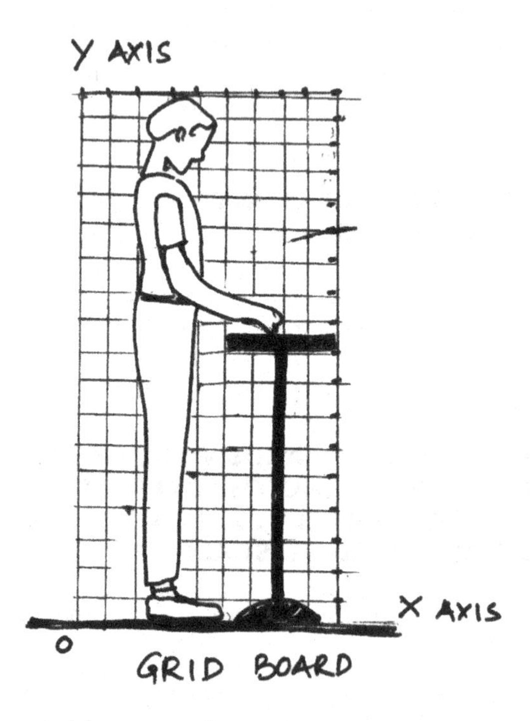

Figure 9.1 Grid board with a mannequin.

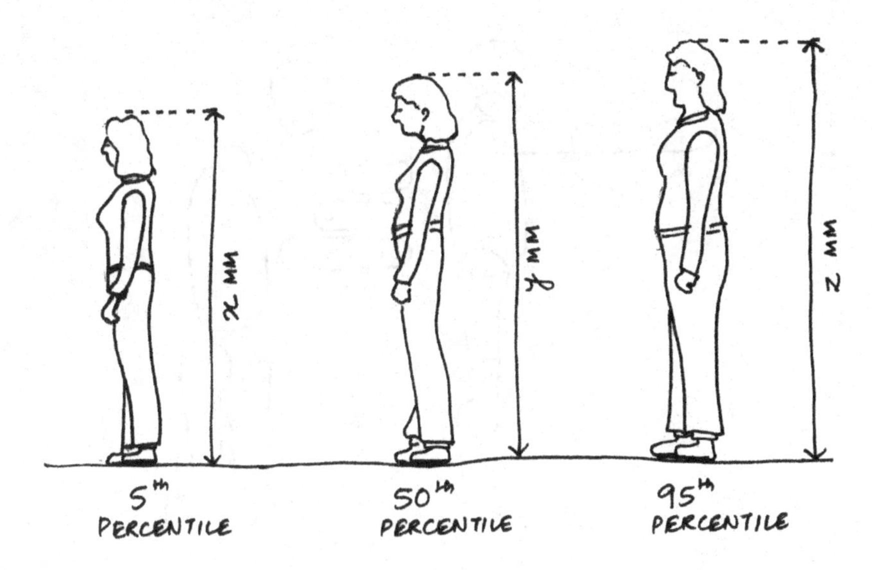

Figure 9.2 Mannequins of different percentile values.

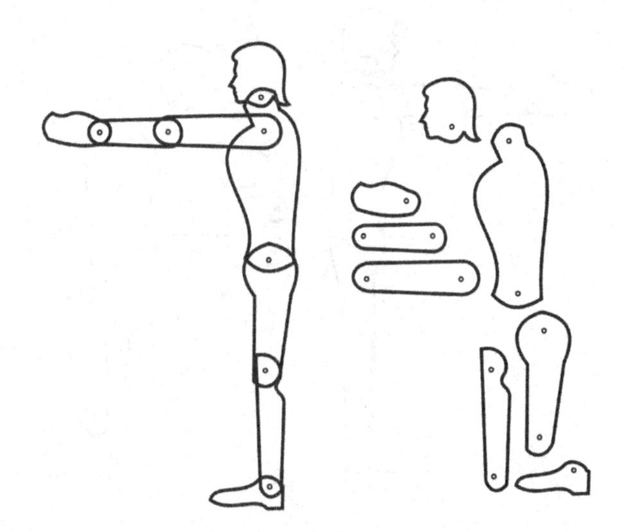

Figure 9.3 Mannequin representing 5th percentile female with different body segments.

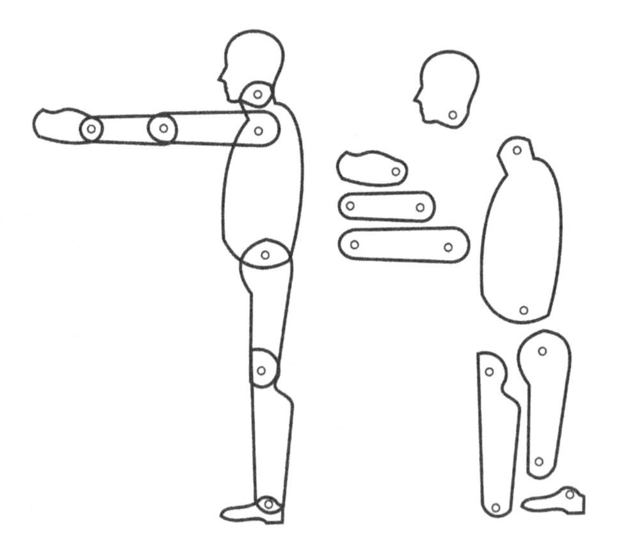

Figure 9.4 Mannequin representing 5th percentile male with different body segments.

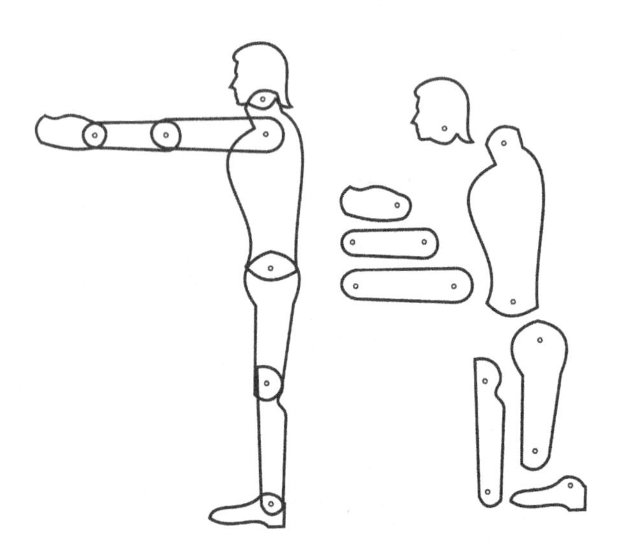

Figure 9.5 Mannequin representing 95th percentile female with different body segments.

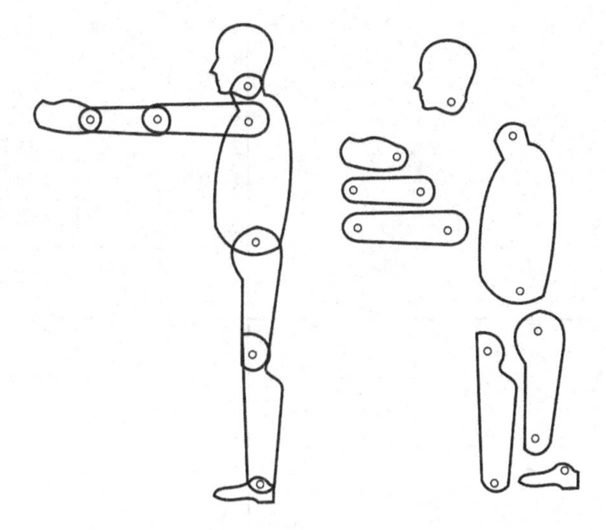

Figure 9.6 Mannequin representing 95th percentile male with different body segments.

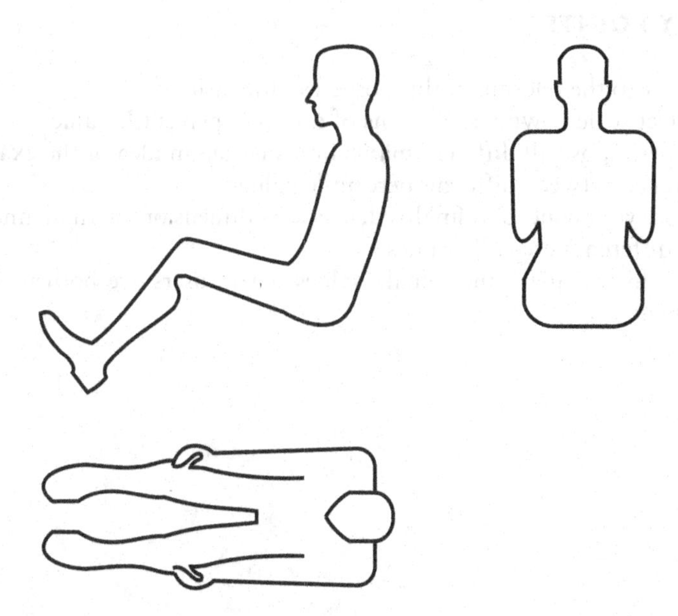

Figure 9.7 Representative mannequins in different views.

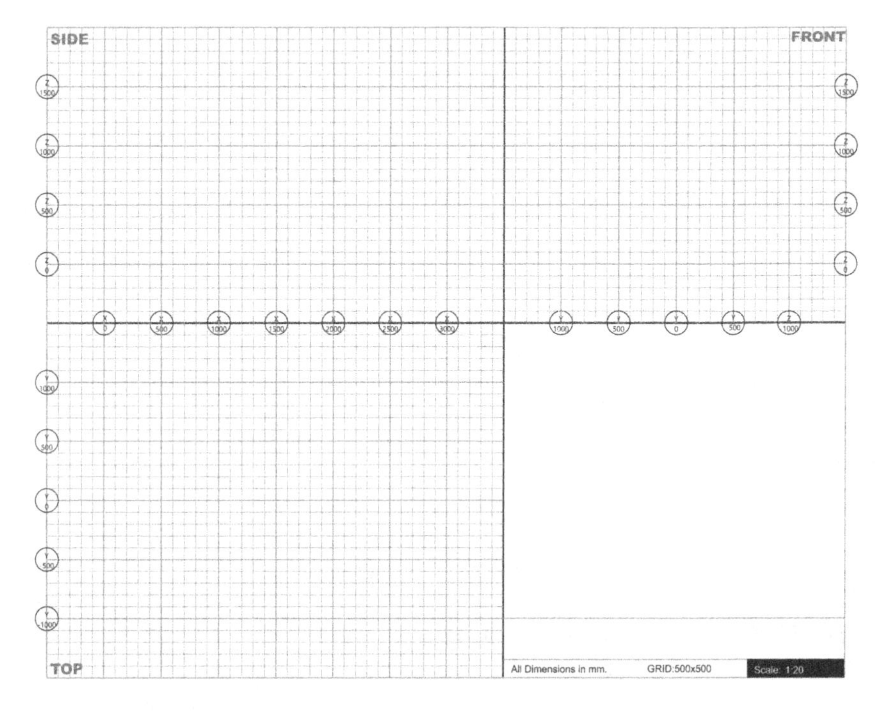

Figure 9.8 Grid board ready to use for workspace ergonomics.

9.5 KEY POINTS

i) Refer to the relevant anthropometric dimensions
ii) Select scale down mannequins of different percentile values
iii) Superimpose all different mannequins to get an idea of the exact difference between different percentile values
iv) Once your concept is finalized, check its dimensions with mannequins of different percentile values
v) Do this for male and female values where users are both male and female

Bibliography

Bridger, R. (2008). *Introduction to Ergonomics*. CRC Press.
Chakrabarti, D. (1997). *Indian Anthropometric Dimensions for Ergonomic Design Practice*. National Institute of Design.
Corlett, E. N. (1973). Human factors in the design of manufacturing systems. *Human Factors*, 15(2), 105–110.
Karwowski, W. (2006). *International Encyclopedia of Ergonomics and Human Factors* (3 Vol. Set). CRC Press.
Kroemer, A. D., & Kroemer, K. H. (2016). *Office Ergonomics: Ease and Efficiency at Work*. CRC Press.
Kroemer, K. H. (2005). *'Extra-Ordinary' Ergonomics: How to Accommodate Small and Big Persons, The Disabled and Elderly, Expectant Mothers, and Children* (Vol. 4). CRC Press.
Kroemer, K. H., & Kroemer, H. J. (1997). *Engineering Physiology: Bases of Human.*, Academic Press
MacLeod, D. (1994). *The Ergonomics Edge: Improving Safety, Quality, and Productivity*. John Wiley & Sons.
Marras, W. S., & Karwowski, W. (2006). *Fundamentals and Assessment Tools for Occupational Ergonomics*. CRC Press.
Mukhopadhyay, P. (2006). Global ergonomics. *Ergonomics in Design*, 14(3), 4–35.
Mukhopadhyay, P. (2008). Time of day effect on performance at a user interface. *Multi: The Journal of Plurality and Diversity in Design*, 1(2)., 36–44
Mukhopadhyay, P. (2009). Ergonomic design of head gear for use by rural youths in summer. *Work*, 34(4), 431–438.
Mukhopadhyay, P. (2013a). Ergonomic design intervention at an Interactive Voice Response (IVR) system in a developing country. *Information Design Journal (IDJ)*, 20(2).148-160.
Mukhopadhyay, P. (2013b). Ergonomic design issues in icons used in digital cameras in India. *International Journal of Art, Culture and Design Technologies (IJACDT)*, 3(2), 51–62.
Mukhopadhyay, P. (2017). Investigation of ergonomic risk factors in snacks manufacturing in central India: Ergonomics in unorganized sector. In *Handbook of Research on Human Factors in Contemporary Workforce Development* (pp. 425–449). IGI Global.

Mukhopadhyay, P. (2019). *Ergonomics for the Layman: Applications in Design.* CRC Press.

Mukhopadhyay, P. (2007.). Heat stress in industry. *On Occupational Health and Safety*, 13.12–13

Mukhopadhyay, P., & Ghosal, S. (2008). Ergonomic design intervention in manual incense sticks manufacturing. *The Design Journal*, 11(1), 65–80.

Mukhopadhyay, P., & Khan, A. (2015). The evaluation of ergonomic risk factors among meat cutters working in Jabalpur, India. *International Journal of Occupational and Environmental Health*, 21(3), 192–198.

Mukhopadhyay, P., & Srivastava, S. (2010a). Ergonomic design issues in some craft sectors of Jaipur. *The Design Journal*, 13(1), 99–124.

Mukhopadhyay, P., & Srivastava, S. (2010b). Ergonomics risk factors in some craft sectors of Jaipur. *HFESA Journal, Ergonomics Australia*, 24(1), 4–14.

Mukhopadhyay, P., & Srivastava, S. (2010c). Evaluating ergonomic risk factors in non-regulated stone carving units of Jaipur. *Work*, 35(1), 87–99.

Mukhopadhyay, P., & Vinzuda, V. (2019). Ergonomic design of a driver training simulator for rural India. In *Advanced Methodologies and Technologies in Artificial Intelligence, Computer Simulation, and Human-Computer Interaction* (pp. 293–311). IGI Global.

Mukhopadhyay, P., Jhodkar, D., & Kumar, P. (2015). Ergonomic risk factors in bicycle repairing units at Jabalpur. *Work*, 51(2), 245–254.

Mukhopadhyay, P., Kaur, J., Kaur, L., Arvind, A., Kajabaje, M., Mann, J., ... & Chakravarty, S. (2013). Ergonomic design analysis of some road signs in India. *Information Design Journal (IDJ)*, 20(3) 220–227

Mukhopadhyay, P., Vinzuda, V., Dombale, S., & Deshmukh, B. (2016). Ergonomic analysis and design of the console panel of a bus rapid transit system in a developing country. *The Design Journal*, 19(4), 565–583.

Mukhopadhyay, P., Vinzuda, V., Naik, S., Karthikeyan, V., & Kumar, P. (2014). Ergonomic analysis of a horse-drawn carriage used for a joy ride in India. *Journal of Human Ergology*, 43(1), 29–39.

Mukhopadhyay, P., Vinzuda, V., Sriram, R., & Doiphode, A. (2012). Ergonomic analysis of a traditional vehicle plying in rural and semi-urban areas in western India. *Journal of Human Ergology*, 41(1–2), 83–94.

Ng, A. W., & Chan, A. H. (2018). Color associations among designers and non-designers for common warning and operation concepts. *Applied Ergonomics*, 70, 18–25.

Panero, J., & Zelnik, M. (1979). *Human Dimension & Interior Space: A Source Book of Design Reference Standards.* Watson-Guptill.

Pheasant, S. (2014). *Bodyspace: Anthropometry, Ergonomics and the Design of Work: Anthropometry, Ergonomics and the Design of Work.* CRC Press.

Pheasant, S., & O'neill, D. (1975). Performance in gripping and turning: A study in hand/handle effectiveness. *Applied Ergonomics*, 6(4), 205–208.

Sanders, M. S., & McCormick, E. J. (1987). *Human Factors in Engineering and Design.* McGRAW-HILL Book Company.

Index

D

F

G

H

L

M

P

R

Printed in the United States
by Baker & Taylor Publisher Services